Generalized Functions and Fourier Analysis

An Introduction

MATHEMATICS LECTURE NOTE SERIES

Published

J. Frank Adams	Lectures on Lie Groups, 1969
E. Artin and J. Tate	Class Field Theory, 1967
Michael F. Atiyah	K-Theory, 1967
Jacob Barshay	Topics in Ring Theory, 1969
Hyman Bass	Algebraic K-Theory, 1968
Melvyn Berger and Marion Berger	Perspectives in Nonlinearity: An Introduction to Nonlinear Analysis, 1968
Armand Borel	Linear Algebraic Groups, 1969
Raoul Bott	Lectures on K(X), 1969
Andrew Browder	Introduction to Function Algebras, 1969
John L. Challifour	Generalized Functions and Fourier Analysis, 1972
Gustave Choquet	Lectures on Analysis Volume I. Integration and Topological Vector Spaces, 1969 Volume II. Representation Theory, 1969 Volume III. Infinite Dimensional Measures and Problem Solutions, 1969
Paul J. Cohen	Set Theory and the Continuum Hypothesis, 1966
Eldon Dyer	Cohomology Theories, 1969
Robert Ellis	Lectures on Topological Dynamics, 1969
Walter Feit	Characters of Finite Groups, 1967
John Fogarty	Invariant Theory, 1969
William Fulton	Algebraic Curves: An Introduction to Algebraic Geometry, 1969
Marvin J. Greenberg	Lectures on Algebraic Topology, 1967
Marvin J. Greenberg	Lectures on Forms in Many Variables, 1969
Robin Hartshorne	Foundations of Projective Geometry, 1967
Robert Hermann	Fourier Analysis on Groups and Partial Wave Analysis, 1969
Robert Hermann	Lectures in Mathematical Physics Volume I, 1970 Volume II, 1972
Robert Hermann	Lie Algebras and Quantum Mechanics, 1970

J.F.P. Hudson	Piecewise Linear Topology, 1969
Irving Kaplansky	Rings of Operators, 1968
Kenneth M. Kapp and Hans Schneider	Completely O-Simple Semigroups: An Abstract Treatment of the Lattice of Congruences, 1969
Joseph B. Keller and Stuart Antman (Editors)	Bifurcation Theory and Nonlinear Eigenvalue Problems, 1969
Irwin Kra	Automorphic Forms and Kleinian Groups, 1972
Serge Lang	Algebraic Functions, 1965
Serge Lang	Rapport sur la cohomologie des groupes, 1966
Ottmar Loos	Symmetric Spaces Volume I. General Theory, 1969 Volume II. Compact Spaces and Classifications, 1969
I.G. Macdonald	Algebraic Geometry: Introduction to Schemes, 1968
George W. Mackey	Induced Representations of Groups and Quantum Mechanics, 1968
Hideyuki Matsumura	Commutative Algebra, 1969
Richard K. Miller	Nonlinear Volterra Integral Equations, 1971
Andrew Ogg	Modular Forms and Dirichlet Series, 1969
Richard S. Palais	Foundations of Global Non-Linear Analysis, 1968
William Parry	Entropy and Generators in Ergodic Theory, 1969
Donald Passman	Permutation Groups, 1968
Walter Rudin	Function Theory in Polydiscs, 1969
David Russell	Optimization Theory, 1970
Gerald E. Sacks	Saturated Model Theory, 1972
Jean-Pierre Serre	Abelian l-Adic Representations and Elliptic Curves, 1968
Jean-Pierre Serre	Algèbres de Lie semi-simples complexes, 1966
Jean-Pierre Serre	Lie Algebras and Lie Groups, 1965
Shlomo Sternberg	Celestial Mechanics Part I, 1969
Shlomo Sternberg	Celestial Mechanics Part II, 1969
Moss E. Sweedler	Hopf Algebras, 1969

Generalized Functions and Fourier Analysis

An Introduction

JOHN L. CHALLIFOUR
Indiana University

1972
W. A. Benjamin, Inc.
ADVANCED BOOK PROGRAM
Reading, Massachusetts

Library of Congress Cataloging in Publication Data

Challifour, John L. 1939
Generalized Functions and Fourier Analysis

(Mathematics lecture note series)
Bibliography: p.
1. Distributions, Theory of (Functional analysis)
2. Differential operators. 3. Fourier series.
4. Fourier transformations. I. Title.
QA324.C45 515'.7 72-8418
ISBN 0-805-31875-5
ISBN 0-805-31876-3 (pbk.)

Reproduced by W. A. Benjamin, Inc., Advanced Book Program, Reading, Massachusetts, from camera-ready copy prepared by the author.

American Mathematical Society (MOS) Subject Classification Scheme (1970): 46F05, 46F10, 42A68

Published simultaneously in Canada.

Manufactured in the United States of America

CONTENTS

PREFACE

The interplay between the development of mathematical theories and the need for better mathematical models in physics has provided fertile ground for both disciplines. This is particularly true with the theory of generalized functions where the need for an extension of the idea of a function dates back to the work of O. Heaviside (1893) on electrical circuits and the studies by P. A. M. Dirac (1926) in relativistic quantum mechanics. Since that time, with the work of many mathematicians, most notably L. Schwartz (1945) and S. L. Sobolev (1938), the theory of generalized functions has become a sophisticated and developed field of mathematics with applications to many areas of analysis, applied mathematics, and theoretical physics. It is the objective of this book to provide a first exposure to this development and several of its applications. Our choice of subject matter has grown from two different teaching experiences. The first three chapters follow a general development of the simplest test function spaces and their related generalized functions up to the point that adequate background is available for the use of Fourier methods. Countable families of norms [2] for the description of convergence for test functions are preferred to seminorms [3], as my experience has been that beginning physics students are more comfortable with the idea of a norm. An early version of the more accessible parts of Chapter two and the beginning of Chapter three formed part of an undergraduate course in advanced calculus at Princeton University during the years 1963-66. Since then, the material has blossomed into its present form which, in recent years, has provided some students in the mathematical physics program at Indiana University with an introduction to the more mathematical aspects of generalized functions. The treatment of invariance with respect to the principal groups of physics in Chapter four has interest in its own right, as well as laying the framework for studying important properties for the equations of mathematical physics in the last chapter. Most sections have complements in the form of simple exercises which may be worked while reading the text. Preparation for the material presented here may be acquired through undergraduate courses in advanced calculus, linear algebra, and complex variables, with some background in general topological notions for Euclidean space.

It is a pleasure to thank Andrew Lenard for his thoughtful comments on Chapter one, and Morton Lowengrub for his reading of the manuscript and for sharing his experiences in teaching similar material. On many occasions my wife, Amber, gave encouragement as well as rescue in my use of the English language. Robert Hutchinson provided aid in both preparation and expert typing of the manuscript.

Bloomington, Indiana — John L. Challifour

August, 1972

1. NORMED LINEAR SPACES

This first chapter serves as a brief introduction to ideas from the theory of normed linear spaces. Given the notion of a linear vector space, topological concepts such as convergence and continuity are introduced by means of a norm, or a countable family of norms. The spaces of test functions, which will define the simplest generalized functions, fall within such a framework.

1.1 LINEAR SPACES

In the following we consider an arbitrary set V whose elements will be denoted by $u, v, w, \ldots$

DEFINITION 1.1. A set V is a linear space over the complex numbers if the operations of addition and multiplication by complex numbers are defined and satisfy the following properties

(i) $u + v = v + u$

(ii) $(u + v) + w = u + (v + w) = u + v + w$

(iii) for any $u \in V$, there exists a unique element $0 \in V$ such that $u + 0 = u$

(iv) for each $u \in V$, there exists a unique element $-u \in V$ such that $u + (-u) = 0$

(v) $\alpha u + \beta v \in V$

(vi) $(\alpha+\beta)u = \alpha u + \beta v$

(vii) $\alpha(u+v) = \alpha u + \alpha v$

(viii) $\alpha(\beta u) = (\alpha\beta)u$

(ix) $(1)u = u$

in which u,v,w are any elements of V and α,β are complex numbers. The elements of a linear space are usually called vectors in analogy with ordinary Euclidean space. By restricting α,β to be real, one defines a linear space over the real numbers.

EXAMPLES. (a) Denote the real numbers by R and the complex numbers by $\mathbb{C}$. These sets are both examples of linear spaces with the ordinary addition and multiplication of real, or complex numbers, respectively.

(b) An extension of the first example is provided by Euclidean n-space, denoted by R^n. A vector $x \in R^n$ is an n-tuple of real numbers $x = (x_1,x_2,\ldots,x_n)$ with operations between vectors defined component-wise

$$x+y = (x_1+y_1,x_2+y_2,\ldots,x_n+y_n) \quad a\,x = (ax_1, ax_2,\ldots,ax_n)$$

for any $x,y \in R^n$ and real numbers a.

In a similar manner, the set $\mathbb{C}^n$ of n-tuples of complex numbers $z = (z_1,z_2,\ldots,z_n)$ becomes a linear space with respect to the operations

$$z+z' = (z_1+z_1', z_2+z_2',\ldots, z_n+z_n') \quad \alpha z = (\alpha z_1, \alpha z_2,\ldots, \alpha z_n)$$

where z, z' are vectors in $\mathbb{C}^n$ and α is a complex number.

(c) A less obvious example would be the set $M(n,\mathbb{C})$ of $n \times n$ matrices with complex entries. This is a linear space under the operations of matrix addition and multiplication by complex numbers

$$A + B = [\alpha_{ij} + \beta_{ij}] \qquad \alpha A = [\alpha\, \alpha_{ij}]$$

for matrices A and B with entries α_{ij} and β_{ij}, respectively. Again, α is any complex number. The subset of $M(n,\mathbb{C})$ consisting of those matrices with non-zero determinant, denoted by $GL(n,\mathbb{C})$, frequently appears in applications to problems in theoretical physics. Important examples are studied in Chapter four.

(d) Let $C^o(R)$ denote the set of continuous functions defined on the real line which take complex values. This set is also a linear space with respect to pointwise operations

$$(f+g)(x) = f(x) + g(x) \qquad (\alpha f)(x) = \alpha\, f(x)$$

for any two functions f,g in $C^o(R)$ and complex number α. These same operations apply to the set $C^m(R)$ of functions on R which take complex values and have continuous derivatives up to order m. $C^m(R)$ is obviously a linear space over the complex numbers.

The next example of a linear space will be most important to our future work. Consider functions, $f(x) = f(x_1,x_2,\ldots,x_n)$, defined on R^n taking complex values. The collection of such functions which possess continuous partial derivatives up to order m will be denoted by $C^m(R^n)$. If m may be taken arbitrarily large, let us write $C^{\infty}(R^n)$ for the resulting linear space and occasionally delete the symbol R^n when this is clear from the context. Our main concern in this chapter is to place appropriate topologies on subsets of C^{∞} (see section 1.7).

When dealing with $C^{\infty}(R^n)$, it is convenient to employ a multi-index notation with a seperate index for each coordinate in R^n. Let $r = (r_1, r_2, \ldots, r_n)$ be an n-tuple of nonnegative integers, and put $|r| = r_1 + r_2 + \ldots + r_n$. Products and derivatives may be written compactly in the same form as for one dimension by means of this notation; for example,

$$x^r = x_1^{r_1} x_2^{r_2} \ldots x_n^{r_n} \qquad D^r f(x) = \partial^{|r|} f(x)/\partial x_1^{r_1} \ldots \partial x_n^{r_n}$$

and

$$x^r y^q = x_1^{r_1} y_1^{q_1} \ldots x_n^{r_n} y_n^{q_n} \qquad D^r D^q f(x) = D^{r+q} f(x) .$$

When factorial expressions appear, the notation $r! = r_1! \, r_2! \ldots r_n!$ is very useful. The context in which this symbol occurs usually makes it clear whether a single or multi-index is being used.

1.2 LINEAR FUNCTIONALS

Given two linear spaces, V and V', a mapping T of V into V', such that

$$T(\alpha u + \beta v) = \alpha \, T(u) + \beta \, T(v)$$

for all $u, v \in V$ and complex numbers α, β, is called linear. Our notation is such that $T(u)$ is the image in V' of the vector u from V. The set of vectors in V', which may be written $u' = T(u)$ for some u in V, is called the image or range of T, denoted $\operatorname{Im} T$. Generally the image is a proper subset, but when $V' = \operatorname{Im} T$, we shall say that T is surjective (onto). In the following, all mappings are restricted to be single valued; namely, there is only one image in V' for each u in V. Further, when $T(u_1) = T(u_2)$ implies $u_1 = u_2$, T is said to be injective (one to one). A mapping which is both injective and surjective is called a bijection from V onto V'. A particular case for such mappings arises when V' is the complex numbers in which case T is called a linear functional on V.

Consider the set of linear mappings from one linear space V to another V'. This set, denoted by $L(V,V')$, is a linear space in its own right with respect to operations

$$(T_1 + T_2)(u) = T_1(u) + T_2(u) \qquad (\alpha T)(u) = \alpha T(u)$$

between linear mappings T_1 and T_2. The collection of linear functionals is just $L(V,\mathbf{C})$.

EXAMPLES. (a) Let $A \in M(n,\mathbf{C})$ with $n > 1$. Then the map $T(A)$ = determinant of A is not a linear functional as is easily seen by taking A to be a suitable multiple of the unit matrix.

(b) Let $[0,1]$ denote the closed unit interval and consider $f \in C^0([0,1])$. The mapping $T_1(f) = \int_0^1 f(x)\,dx$ is a linear functional. Is this true if we replace $f(x)$ by $|f(x)|$ in this expression?

One may also consider a linear mapping T between vectors in V and a linear space V' which is not necessarily defined on all elements of V. For such a situation, the set of those vectors for which $T(u)$ is defined will be called the domain of T, written $D(T)$. The differential operator, one of the most important linear operators in mathematical physics, provides a ready example. More precisely, consider the linear space $C^0(R^n)$ and define an operator

$$(D_i f)(x) = \partial f(x)/\partial x_i \qquad 1 \le i \le n .$$

Clearly $D_i = \partial/\partial x_i$ is defined only on the differentiable functions in $C^0(R^n)$ and on that domain determines a linear operator. The operator D^r introduced at the end of section 1.1 determines a linear operator on a domain consisting of the r-times differentiable functions.

EXAMPLES. (c) An example of a linear functional which is not everywhere defined is given by $T(f) = \int_0^1 f(x)/x \, dx$ with $f \in C^0([0,1])$. In order that $T(f)$ be finite, f must vanish sufficiently rapidly as $x \to 0^+$ so that the integral exists. In particular, T is defined on all f which behave like x^α as $x \to 0^+$, with $\alpha > 0$. T is certainly not defined on functions which take constant nonzero values at the origin.

PROBLEMS. (1) Is the function defined as follows in the domain of

d/dx ? Put $f(x) = 0$ for $0 \leq x \leq 1/2$ and $\exp(-1/(x-1/2))$ for $1/2 < x \leq 1$.

(2) Consider the following linear mappings on $C^o([0,1])$

$$(T_1 f)(x) = d/dx \int_0^x dt\, f(t) \qquad (T_2 f)(x) = \int_0^x dt\, df/dt \ .$$

Discuss their respective domains of definition and how they relate to the Fundamental Theorem of Calculus.

1.3 NORMED LINEAR SPACES AND TOPOLOGIES

The concept which will enable us to introduce a topology into a linear space is that of a norm, defined as follows.

DEFINITION 1.2. A linear space V is said to be normed if to each $u \in V$ there is associated a non-negative number $\|u\|$, such that

(i) $\|u+v\| \leq \|u\| + \|v\|$

(ii) $\|\alpha u\| = |\alpha|\ \|u\|$ for any complex α

(iii) $\|u\| = 0$ implies $u = 0$.

This number $\|u\|$ is called the norm of u and V is said to be a normed space.

PROBLEMS. (1) Show that R^n is a normed space with respect to either of the norms

$$\|x\| = \Big(\sum_{i=1}^{n} x_i^2\Big)^{\frac{1}{2}} \qquad \|x\|' = \sup_{1\leq i\leq n} |x_i| \ . \tag{1.1}$$

(2) Consider the space $C^m([0,1])$. Show that this may

be normed with respect to any of the norms $\|f\|_r = \sup_{x,q\le r} |f^{(q)}(x)|$, $0 \le r \le m$, in which $f^{(r)}$ denotes the r-th derivative. How are these norms related? If we take $0 < r \le m$, is $\|f\|_r$ still a norm?

(3) Let U be the interior of the unit disc in the complex plane, and by $H(U)$, denote the set of functions analytic in U . Is $H(U)$ a normed linear space with respect to the norms $\|f\|_r = \sup_{z\in U, 0\le q\le r} |f^{(q)}(z)|$, where $f \in H(U)$? Suppose γ is any simple, closed, continuous curve in U . Show that $\int_\gamma dz\, f(z)/z$ is a linear functional on $H(U)$ and determine its value.

As remarked at the beginning of the chapter, there is a natural way to introduce a topology into a normed linear space. As we shall need more general notions in this respect, let us recall some facts from point set topology. A topology on a set X is a collection τ of subsets of X such that

(i) the empty set $\emptyset$ and $X \in \tau$

(ii) if $U_i \in \tau$ for $i=1,2,\dots,n$ then $\bigcap_{i=1}^{n} U_i \in \tau$ (1.2)

(iii) the union of arbitrary collections of members from τ is again an element of τ .

Under these conditions, X is said to be a topological space with respect to the topology τ . Members of τ are called the open sets for X . It could well happen that the set X admits two topologies, τ_1 and τ_2 . In this case, τ_1 is said to be weaker (smaller) than τ_2 if $\tau_1 \subseteq \tau_2$. When either topology is weaker than the other, we say that they are equivalent topologies for X .

In practice, it may be rather hard to specifically exhibit a collection of subsets of a given set which satisfy (1.2), and for this reason, the idea of a base for a topology is useful. A collection σ of subsets of X is called a base for a topology τ if

(i) $\sigma \subseteq \tau$

(ii) every set in τ is a union of sets in σ. (1.3)

The question of when a family of subsets form a base is settled in the next theorem.

THEOREM 1.3. Let σ be a collection of subsets of a set X and denote the collection of all unions of sets in σ by τ. Then τ is a topology for X with base σ, if and only if, the following two conditions hold

(a) X = union of sets in σ

(b) for $U_1, U_2 \in \sigma$ and $x \in U_1 \cap U_2$, there exists $U_3 \in \sigma$ such that $x \in U_3 \subseteq U_1 \cap U_2$.

Proof. If σ is a base for τ, then X is a union of sets in σ. Moreover $\sigma \subseteq \tau$ implies $U_1 \cap U_2 \in \tau$; hence, if x is as in part (b) of the theorem, there is a member of σ containing x and contained in $U_1 \cap U_2$. The conditions are then necessary.

Suppose σ satisfies the stated hypotheses. Clearly τ has the properties (i), (ii) of (1.2). Let $U_1 \cap U_2$ be an intersection of sets in τ containing x. There exist sets $U_3 \subseteq U_1$, $U_4 \subseteq U_2$ from σ for which $x \in U_3 \cap U_4$. By property (b), we may find a $U_5 \in \sigma$ with $x \in U_5 \subseteq U_3 \cap U_4 \subseteq U_1 \cap U_2$, from which we conclude that $U_1 \cap U_2$ is a union of sets from σ; thus, in τ. The conditions are also sufficient.

For a normed space, Theorem 1.3 is applied in the following manner. To define a system of open sets for V, consider $u_o \in V$ and any real number $\epsilon > 0$. An open ϵ-neighborhood of u_o is the set

$$N_\epsilon(u_o) = \{u \in V \mid \|u - u_o\| < \epsilon\} \ . \tag{1.4}$$

A closed neighborhood, $\bar{N}_\epsilon(u_o)$, has the inequality replaced by equality. A base for u_o is obtained by taking the collection of neighborhoods $N_\epsilon(u_o)$ as ϵ varies over all positive numbers. Due to the linear property of V, if $N_\epsilon(0)$ is an ϵ-neighborhood of 0, then $u_o + N_\epsilon(0)$ is an ϵ-neighborhood of u_o. A topology for V then results by taking a base which is the collection of all such neighborhoods for each point. The resulting topology, induced on V in this way, is called the norm topology.

It may well happen that a linear space V admits more than one norm, and consequently, could be regarded as a different topological space for each such norm. Suppose we have two norms where $\| \ \|_1$ leads to a topology τ_1 and $\| \ \|_2$ to τ_2. If $\tau_2 \subseteq \tau_1$, then for every $\epsilon_2 > 0$, there exists an $\epsilon_1 > 0$ such that $\|u\|_2 < \epsilon_2$ implies $\|u\|_1 < \epsilon_1$. This comparison of the two topologies motivates the following comparison of the two norms.

DEFINITION 1.4. Let V be a linear space with two norms $\| \ \|_1$, $\| \ \|_2$. We say that $\| \ \|_1$ is weaker than $\| \ \|_2$ if $\| u \|_1 \leq c\| u \|_2$ holds for all $u \in V$.

One could equally well say that norm $\| \|_2$ is stronger than the norm $\| \|_1$. It is important in the definition that the constant c may be chosen independently of the vectors in V . It is an unfortunate choice of terminology that has stronger norms inducing weaker topologies. Two norms will be considered equivalent if they are both stronger and weaker than each other. The same is true for their respective topologies.

EXAMPLE. (a) R^n is normed with respect to either of the norms in (1.1). From the inequalities

$$\left(\sum_{i=1}^{n} x_i^2\right)^{\frac{1}{2}} \le n^{\frac{1}{2}} \sup_{1\le i\le n} \|x_i\| \le n\left(\sum_{i=1}^{n} x_i^2\right)^{\frac{1}{2}}$$

we find that $\|x\| \le n^{\frac{1}{2}} \|x\|' \le n\|x\|$; hence, that $\| \|$ and $\| \|'$ induce equivalent topologies on R^n .

PROBLEMS. (4) Let V be any normed space and define $d(u_1,u_2) = \|u_1 - u_2\|$ for any two vectors in V . Show that

(i) $d(u_1,u_2) + d(u_1,u_3) \ge d(u_2,u_3)$

(ii) $d(u_1,u_2) \ge 0$

(iii) $d(u_1,u_2) = 0$ if, and only if, $u_1 = u_2$.

A function, $d(u,0)$ on V , with these properties is called a metric. A linear space which admits a metric is called a metric space.

(5) Suppose $\| \|_1$, $\| \|_2$ are two distinct norms on a linear space V . Are $\|u\| = \max[\|u\|_1, \|u\|_2]$ and $\|u\|' = \min[\|u\|_1, \|u\|_2]$ norms on V ? What is their relationship to

the original norms?

1.4 COMPLETE NORMED SPACES

Given a normed space V, limiting processes are defined using the norm topology.

DEFINITION 1.5. A sequence $\{u_n\}_{n=1}^{\infty}$ of vectors in V converges to $u \in V$ if $\lim_{n\to\infty} \|u - u_n\| = 0$.

An alternative statement of this idea is to say that for every $\epsilon > 0$ there exists $n_o(\epsilon)$ such that $u_n \in N_\epsilon(u)$ whenever $n \geq n_o$. This is the familiar definition of convergence for a topological space. As for the real and complex numbers, we have the idea of Cauchy convergent sequences.

DEFINITION 1.6. A sequence $\{u_n\}_{n=1}^{\infty}$ is said to be Cauchy if, for all $\epsilon > 0$, there exists an $n_o(\epsilon)$ such that $\|u_n - u_m\| < \epsilon$ when $n \geq m \geq n_o$.

Clearly every convergent sequence is Cauchy. For suppose $\{u_n\}$ converges to u, then using property (i) of a norm $\|u_n - u_m\| \leq \|u_n - u\| + \|u - u_m\| \leq 2\epsilon$ for m and n taken sufficiently large. It is important to observe that, in general, the converse is not true. When, however, this is the case, we have the idea of a complete normed space.

DEFINITION 1.7. A normed space V is complete if every Cauchy sequence converges to a limit in V . The convergence is with respect to the norm topology.

In the mathematical literature, complete normed spaces are also termed Banach spaces in honor of the Polish mathematician, S. Banach, who first exploited them systematically. It is a simple matter to prove that every finite dimensional linear space is a complete normed space with respect to either of the norms (1.1). One first introduces a fixed basis, then uses completeness of the real or complex numbers (Bolzano-Weierstrass) on the components of a given vector with respect to this basis. This does not extend to normed spaces which cannot be spanned by a finite number of vectors.

PROBLEMS. (1) Let $\{\alpha_n\}$ be a sequence of complex numbers which tend to zero. Show that $\{\alpha_n u\}$ tends to zero in the norm topology.

(2) Let $\{u_n\}$ tend to a vector v in the norm topology. Show that for any vector u , $\{u + u_n\}$ tends to $u + v$.

Even though completeness is a rather special property of a normed space, any normed space V can be embedded in a natural and minimal way in a Banach space $\hat{V}$. The space $\hat{V}$ is called the completion of V and is constructed as follows.

Let $\hat{V}$ denote the collection of Cauchy sequences $\hat{u} = \{u_n\}_{n=1}^{\infty}$ taken from V . In particular, $V \subset \hat{V}$ as the vectors u may be identified with the sequences $\{u,u,\ldots,u,\ldots\}$. This will be understood in the following arguments. Two Cauchy sequences, $\{u_n\}$ and $\{u_n'\}$, will be regarded as equivalent if $\lim_{n\to\infty} \|u_n - u_n'\| = 0$. The distinct elements of $\hat{V}$ are then classes of equivalent Cauchy sequences from V. Now that $\hat{V}$ has been defined, it remains to show that it is a Banach space with respect to some norm, say $|||\cdot|||$, for which if $\hat{u} = \{u,u,\ldots,u,\ldots\}$, then $|||\hat{u}||| = \|u\|$ where $\|\cdot\|$ is the original norm on V. This will be the case if we define

$$|||\hat{u}||| = \lim_{n\to\infty} \|u_n\| \tag{1.5}$$

for any $\hat{u} \in \hat{V}$.

First note that the linear operations in $\hat{V}$ are addition of Cauchy sequences and their multiplication by complex numbers. Second, the expression (1.5) is well defined for every Cauchy sequence as a consequence of the inequality

$$|\|u_n\| - \|u_m\|| \leq \|u_n - u_m\| \tag{1.6}$$

and the completeness of the real numbers. Namely, $\{\|u_n\|\}$ is a Cauchy sequence of real numbers, so the limit in (1.5) exists. To verify that $|||\cdot|||$ is a norm, one readily checks that

(a) $|||\hat{u}+\hat{v}||| = \lim_{n\to\infty} \|u_n + v_n\| \leq \lim_{n\to\infty} \|u_n\| + \lim_{n\to\infty} \|v_n\| = |||u||| + |||v|||$

(b) $|||\alpha u||| = \lim_{n\to\infty} \|\alpha u_n\| = |\alpha| \lim_{n\to\infty} \|u_n\| = |\alpha|\ |||u|||$

(c) $|||u||| = 0$ implies $\lim_{n\to\infty} \|u_n\| = 0$ and $\hat{u}$ is equivalent to $\hat{0} = \{0,\ldots,0,\ldots\}$, the zero vector in $\hat{V}$.

$\hat{V}$ is complete with respect to $|||\cdot|||$. For let $\{\hat{u}_p\}$ be a Cauchy sequence in $\hat{V}$, where for each member of the sequence, the corresponding Cauchy sequence in V is written

$\hat{u}_p = \{u_{p1}, u_{p2}, \ldots, u_{pn}, \ldots\}$. Since each $\hat{u}_p$ is Cauchy, we deduce

$$\|u_{pm} - u_{pn}\| < 1/p \quad \text{if} \quad m \geq n \geq \chi(p) \tag{1.7}$$

for $\chi(p)$ large enough. This in turn implies that

$$\lim_{m\to\infty} \|u_{pm} - u_{pn}\| = |||\hat{u}_p - u_{pn}||| < 1/p \quad \text{if} \quad n \geq \chi(p) \ . \tag{1.8}$$

We may then approximate $\hat{u}_p$ in the $|||\cdot|||$-norm by the sequence $\{u_{pn}, u_{pn}, \ldots\}$ for n large enough depending upon p . Consider the "diagonal" sequence $\hat{u} = \{u_{1\chi(1)}, u_{2\chi(2)}, \ldots, u_{n\chi(n)}, \ldots\}$. For this

$$\begin{aligned} \|u_{m\chi(m)} - u_{n\chi(n)}\| &\leq |||u_{m\chi(m)} - \hat{u}_m||| + |||\hat{u}_m - \hat{u}_n||| + |||\hat{u}_n - u_{n\chi(n)}||| \\ &\leq |||\hat{u}_m - \hat{u}_n||| + 1/m + 1/n \end{aligned} \tag{1.9}$$

and the right-hand side may be made arbitrarily small by sufficiently large m,n . The diagonal sequence is then Cauchy in V , and $\hat{u} \in \hat{V}$. We are almost finished since $\{\hat{u}_p\}$ converges to u in the $|||\cdot|||$-norm. To see this, another 2ϵ-argument suffices; namely,

$$\begin{aligned} |||\hat{u}-\hat{u}_p||| &\leq |||\hat{u} - u_{p\chi(p)}||| + |||u_{p\chi(p)} - \hat{u}_p||| \\ &\leq 1/p + \lim_{n\to\infty} \|u_{n\chi(n)} - u_{p\chi(p)}\| \end{aligned} \tag{1.10}$$

which tends to zero as p increases. Summarizing these results, we have

PROPOSITION 1.8. Every normed space V is either complete or is contained in a Banach space $\hat{V}$ without changing the norm of elements in V . Moreover, as a topological space, V is dense in the norm topology of $\hat{V}$.

It is clear from (1.10) that V is dense in $\hat{V}$ as any element $\hat{v} \in \hat{V}$ is the limit of a sequence $\{\hat{u}_n\}$ which, in turn, is given by the diagonal sequence $\hat{u}$ of elements in V . A special case is the situation in which a normed space is separable; that is,

DEFINITION 1.9. A normed space V is separable if it contains a countable subset which is dense with respect to the norm topology.

Consider a linear space V which admits two different norms $\|\cdot\|_1$, $\|\cdot\|_2$ and is not complete with respect to either of them. Suppose, moreover, that the 1-norm is weaker than the 2-norm. Then every sequence Cauchy for the 2-norm is also Cauchy for the 1-norm by virtue of the estimate $\|u_n - u_m\|_1 \leq c\|u_n - u_m\|_2$. If $\hat{V}_1$, $\hat{V}_2$ denote the completions of V with respect to the two norms, one concludes that $\hat{V}_2 \subset \hat{V}_1$. As this inclusion indicates, there is a natural mapping of $\hat{V}_2$ onto $\hat{V}_1$ obtained as follows. Each $\hat{u}_2 \in \hat{V}_2$ is the limit of a Cauchy sequence $\{u_n\}$ in V which, in turn, has a limit $\hat{u}_1 \in \hat{V}_1$. The map $T_{12}\colon V_2 \to V_1$ is defined by $T_{12}(\hat{u}_2) = \hat{u}_1$ and is clearly linear and onto. Generally, as example (a) below illustrates, T_{12} is not injective; namely, distinct elements in $\hat{V}_2$ may be mapped into the same element in $\hat{V}_1$. This difficulty can be avoided by making a further requirement on the two norms which will guarantee that the topologies of $\hat{V}_1$ and $\hat{V}_2$ will be compatible.

DEFINITION 1.10. Two norms $\|\cdot\|_1$, $\|\cdot\|_2$ on a linear space V are said to be compatible if every sequence which is Cauchy with respect to both norms and converges to zero in one norm, also converges to zero in the other norm.

When $\|\cdot\|_1$, $\|\cdot\|_2$ are compatible for V, the mapping T_{12} is injective. For suppose, $\{u_n\}$, $\{u_n'\}$ are two Cauchy sequences from V converging to $\hat{u}$ and $\hat{u}'$, respectively, in $\hat{V}_2$, while in $\hat{V}_1$, they both converge to $\hat{v}$. Then $\{u_n - u_n'\} \to 0$ in $\hat{V}_1$, but to $\hat{u} - \hat{u}'$ in $\hat{V}_2$, contradicting the assumed compatibility of the two norms unless $\hat{u} = \hat{u}'$.

EXAMPLES. (a) Let $V = C^1([0,1])$ with norms

$$\|f\|_1 = \sup_x |f(x)| \qquad \|f\|_2 = \sup_x |f(x)| + |f^{(1)}(0)| .$$

Clearly $\|f\|_1 \le \|f\|_2$ and $\hat{V}_2 \subset \hat{V}_1$. Consider the sequences $\{(1-x)^n/n\}$, $\{(1-x)^n/n^2\}$. These both converge to zero in $\hat{V}_1$, but in $\hat{V}_2$, the first converges to one and the second to zero.

(b) The ℓ_1 space. Consider the collection of infinite sequences $\xi = (\xi_1, \xi_2, \ldots, \xi_n, \ldots)$ of complex numbers ξ_i. Such a collection is a linear space with respect to the operations

$$\alpha\xi = (\alpha\xi_1, \ldots, \alpha\xi_n, \ldots) \qquad \xi + \eta = (\xi_1 + \eta_1, \ldots, \xi_n + \eta_n, \ldots)$$

and is a natural extension of the finite dimensional case $\mathbb{C}^n$ (see example 1.1 (b)).

Let ℓ_1 denote the normed space obtained by considering only those sequences for which

$$\| \xi \|_1 = \sum_{n=1}^{\infty} | \xi_n | < \infty \tag{1.11}$$

Then we prove that ℓ_1 is an infinite dimensional, separable, Banach space.

(i) ℓ_1 is clearly infinite dimensional in the sense that the vectors of the form $e_n = (0,\ldots,0,1,0,\ldots)$, with 1 in the n-th place, are linearly independent for all positive integers n .

(ii) Separability for ℓ_1 will follow from the fact that the complex rationals form a countable dense subset with respect to the norm (1.11). More precisely, let

$$U = \{\xi \in \ell_1 | \ \xi = (\rho_1, \rho_2, \ldots, \rho_n, 0, \ldots) \ \ \rho_n = s_n + it_n \text{ with } s_n, t_n \text{ rational numbers for } 1 \le n < \infty\} .$$

Then U is countable and dense in ℓ_1 . To see this last statement, choose some $\xi \in \ell_1$ and any $\epsilon > 0$. Then there exists $n_o(\epsilon)$ such that $\sum_{n=n_o+1}^{\infty} |\xi_n| < \epsilon/2$ and the vector $\xi' = (\xi_1, \xi_2, \ldots, \xi_{n_o}, 0, \ldots)$ satisfies

$$\|\xi - \xi'\|_1 < \epsilon/2 \quad .$$

Now for each component of ξ' , we can find a complex rational ρ_k such that

$$|\xi_k - \rho_k| < \epsilon/2n_o \qquad 1 \le k \le n_o$$

Then $\xi'' = (\rho_1, \rho_2, \ldots, \rho_{n_o}, 0, \ldots) \in U$ with

$$\|\xi' - \xi''\|_1 = \sum_{n=1}^{n_o} |\xi_n - \rho_n| < \epsilon/2$$

and the triangle inequality gives

$$\|\xi - \xi''\|_1 \le \|\xi - \xi'\|_1 + \|\xi' - \xi''\|_1 < \epsilon \ .$$

(iii) Finally, we deal with the completeness of ℓ_1. Choose any Cauchy sequence, say $\{\xi_n\}$, and for each vector write its components $\xi_p = (\xi_{p1},\ldots,\xi_{pk},\ldots)$. Then, given $\epsilon > 0$, we may select $n_o(\epsilon)$ for which

$$\|\xi_p - \xi_q\|_1 < \epsilon \qquad p \ge q \ge n_o \ .$$

For a particular component this also implies

$$|\xi_{pn} - \xi_{qn}| \le \|\xi_p - \xi_q\|_1 < \epsilon \qquad p \ge q \ge n_o$$

thereby for each $n\{\xi_{pn}\}_{p=1}^{\infty}$ is a Cauchy sequence of complex numbers with a limit $\lim_{p\to\infty} \xi_{pn} = \xi_n'$. Form the vector $\xi' = (\xi_1',\xi_2',\ldots,\xi_n',\ldots)$ whose n-th component is the limit of ξ_{pn} and observe that from the estimate

$$\sum_{k=1}^{m} |\xi_{pk} - \xi_{qk}| \le \|\xi_p - \xi_q\|_1 < \epsilon \qquad p \ge q \ge n_o$$

as $p \to \infty$, we find

$$\sum_{k=1}^{m} |\xi_k' - \xi_{qk}| < \epsilon \qquad q \ge n_o$$

for all positive integers m. As the right-hand side is independent of m, this means

$$\sum_{k=1}^{\infty} |\xi_{qk} - \xi'_k| < \varepsilon \qquad q \geq n_o .$$

Consequently, for q large enough $\{\xi_q - \xi'\} \in \ell_1$, and by the triangle inequality,

$$\| \xi' \|_1 \leq \|\xi' - \xi_q\|_1 + \| \xi_q \|_1 < \varepsilon + \| \xi_q \|_1 .$$

This shows that $\xi' \in \ell_1$ and $\lim_{q \to \infty} \xi_q = \xi'$.

(c) As an example of an incomplete normed space, let V be the set of those sequences in example (b) which have only a finite number of non-zero components. Clearly V is normed with respect to (1.11), it is, however, not complete since the sequence

$$\xi_k = (1, 1/4, 1/9, \ldots, 1/k^2, 0, \ldots)$$

converges as $k \to \infty$ to a vector in ℓ_1 which is certainly not in V. In fact, upon taking the completion with respect to (1.11), a slight modification of part (ii) in example (b) shows that $\hat{V}_1 = \ell_1$.

PROBLEMS. (3) Consider $C^o([0,1])$ with the two norms $\|f\|_1 = \sup_{0 \leq x \leq 1} |f(x)|$, $\|f\|_2 = \int_0^1 dx\, |f(x)|$. Are these compatible?

(4) Consider $C^o([0,1])$ with the norm $\|f\| = \int_0^1 dx\, |f(x)|$. Show that this space is not complete with respect to this norm. (Hint: Approximate a non-continuous function with finite integral by continuous functions.)

(5) Consider the space of complex sequences

$\xi = (\xi_1, \xi_2, \ldots, \xi_n, \ldots)$ introduced in example (b). For any integer $1 \leq p < \infty$, generalize ℓ_1 to the spaces

$$\ell_p = \left\{ \xi \mid \|\xi\|_p = \left(\sum_{n=1}^{\infty} |\xi_n|^p \right)^{1/p} < \infty \right\} . \tag{1.12}$$

Show that ℓ_p is a separable, Banach space. (<u>Hint</u>: To check that (1.12) is a norm, use Minkowski's inequality for positive real numbers

$$\left(\sum_{n=1}^{\infty} (a_n + b_n)^p \right)^{1/p} \leq \left(\sum_{n=1}^{\infty} a_n^p \right)^{1/p} + \left(\sum_{n=1}^{\infty} b_n^p \right)^{1/p}$$

To show completeness, follow example (b).)

(6) Consider the space of sequences in example (b). Show that

$$\ell_\infty = \{ \xi \mid \|\xi\|_\infty = \sup_n |\xi_n| < \infty \} \tag{1.13}$$

is a Banach space which is not separable.

(7) Let $\|\cdot\|_1$, $\|\cdot\|_2$ be two compatible norms for a linear space V. Define $\|\cdot\|_2' = \max [\|\cdot\|_1 , \|\cdot\|_2]$ and show that $\|\cdot\|_1$ and $\|\cdot\|_2'$ are also compatible.

1.5 COUNTABLY NORMED SPACES

Let V be a linear space which becomes a normed space with respect to any one of a countable family of mutually compatible norms $\{\|\cdot\|_i\}_{i=1}^{\infty}$. A priori, there need be no relation between the $\|\cdot\|_i$ other than the topological condition of compatibility. Define now new norms

$$
\begin{aligned}
\|u\|_1' &= \|u\|_1 \\
\|u\|_2' &= \max_{i=1,2}\{\|u\|_i\} \\
&\vdots \\
\|u\|_n' &= \max_{1\le i\le n}\{\|u\|_i\} \\
&\vdots
\end{aligned}
\tag{1.14}
$$

As shown in problem 1.4 (7), the prime norms induce related compatible topologies on V, and the topological properties may be stated accordingly in terms of either set. The advantage of the prime norms lies in the fact that they are ordered by $\|u\|_1' \le \|u\|_2' \le \dots \le \|u\|_n' \le \dots$. Such a family will be said to have increasing strength as their respective norm topologies satisfy $\tau_1' \supseteq \tau_2' \supseteq \dots \supseteq \tau_n' \supseteq \tau_{n+1}' \supseteq \dots$. In future, we shall assume without loss of generality that a countable family $\{\|\cdot\|_n\}_{n=1}^{\infty}$ on V has increasing strength and omit the prime notation.

With this last remark in mind, notice the inclusions

$$V \subset \dots \subset \hat{V}_{n+1} \subset \hat{V}_n \subset \dots \subset \hat{V}_2 \subset \hat{V}_1 . \tag{1.15}$$

The original linear space V becomes a topological space which is dense in each one of the $\hat{V}_n$ and has a topology whose open sets are formed from a base consisting of the neighborhoods

$$N_{\epsilon,n}(0) = \{u \in V \mid \|u\|_n < \epsilon\} \quad n=1,2,\dots \quad \text{all positive } \epsilon . \tag{1.16}$$

It is worthwhile remarking that if the family $\{\|\cdot\|_n\}$ is not ordered according to increasing strength, the basic neighborhoods are the sets

$$N_{\varepsilon,n}(0) = \{u \in V \mid \|u\|_1 < \varepsilon \; \|u\|_2 < \varepsilon \; \ldots \; \|u\|_n < \varepsilon\} . \qquad (1.17)$$

Collecting these comments together, we make a formal definition

DEFINITION 1.11. A linear space V with a countable family of increasing, mutually compatible norms, and a topology defined by the base of sets (1.16), is called a countably normed space.

One of the first questions posed by this definition is the completeness of such spaces. This is answered in our next result.

THEOREM 1.12. A countably normed space V is complete if, and only if, $V = \bigcap_{n=1}^{\infty} \hat{V}_n$.

Proof. Notice that (1.15) implies $\bigcap_{n=1}^{\infty} \hat{V}_n$ exists and contains V .

The condition is sufficient. For assume $V = \bigcap_{n=1}^{\infty} \hat{V}_n$ and $\{u_p\}$ is a Cauchy sequence in V . Then $\{u_p\}$ belongs to each V_n , and by (1.16), is also Cauchy in each of these spaces. As these are complete by construction, for each integer n there exists a $\hat{v}_n \in \hat{V}_n$ such that $\lim_{p\to\infty} \|u_p - \hat{v}_n\|_n = 0$. Recall our remarks preceeding definition 1.10 and conclude that the bijection $T_{n\,n-1}: V_n \to V_{n-1}$ allows us to identify the $\hat{v}_n$ uniquely with a vector u in the intersection of all the $\hat{V}_n$. This vector u has the property that $\lim_{p\to\infty} \|u_p - u\|_n = 0$ for every integer n ; hence V is complete.

The necessity of this condition follows easily. Suppose V is complete and $u \in \bigcap_{n=1}^{\infty} \hat{V}_n$. The fact that V is dense in each $\hat{V}_n$ allows us to find a sequence of vectors $\{u_n\}$ in V for which $\|u - u_n\|_n < 1/n$. The estimate

$$\begin{aligned} \|u_p - u_q\|_n &\le \|u - u_p\|_n + \|u - u_q\|_n \\ &\le \|u - u_p\|_p + \|u - u_q\|_q < 2/q \quad \text{for } p \ge q \ge n \end{aligned} \tag{1.18}$$

allows us to infer that $\{u_n\}$ is Cauchy with respect to the topology (1.16); therefore, by the completeness of V , converges to a vector $u \in V$. Hence, $\bigcap_{n=1}^{\infty} \hat{V}_n \subset V$ and the theorem is demonstrated. Notice that the increasing strength of the norms appeared in (1.18).

The last generalization of a normed space which we shall need is that of an inductive limit. Consider two topological spaces X_1 and X_2 with their respective topologies τ_1 and τ_2 . If the underlying sets stand in the relation $X_1 \subseteq X_2$, it is reasonable to ask what relation might hold between their topologies. In this direction, recall the notion of the relative topology of X_1 . This is the collection of sets in X_1 of the form $X_1 \cap U$ where $U \in \tau_2$. We try to match this with τ_1 as follows

DEFINITION 1.13. Let (X_1, τ_1) , (X_2, τ_2) be two topological spaces for which $X_1 \subseteq X_2$. We say that this inclusion extends to the respective topological spaces if τ_1 agrees with the relative topology of X_1 as a subset of (X_2, τ_2) .

This leads directly to our last concept from the theory of linear spaces

DEFINITION 1.14. Let $W_1 \subset W_2 \subset \dots \subset W_n \subset W_{n+1} \subset \dots$ be an increasing sequence of complete countably normed spaces satisfying definition 1.13. Then the linear space $\bigcup_{\alpha=1}^{\infty} W_\alpha$ is called the strict inductive limit of the family $\{W_\alpha\}$.

In section 1.7 we shall introduce certain test function spaces, the most important of which is a strict inductive limit. Let us close this section by showing that the strict inductive limit is indeed a topological space.

THEOREM 1.15. Let W be the strict inductive limit of the family $\{W_\alpha\}$ and let σ be the collection of sets in W whose intersection with each W_α is an open set of the form (1.16). Then σ is a base for a topology for W.

Proof. Clearly W is a union of sets in σ since each W_α is a union of sets of the form (1.16). The set $\bigcup_{\alpha=1}^{\infty} U_\alpha$, defined as follows, will satisfy the second requirement of Theorem 1.3. Let S_1, S_2 be members of σ with $u \in S_1 \cap S_2$ and put for U_α a suitable neighborhood in $S_1 \cap S_2 \cap W_\alpha$ containing u.

It is worthwhile remarking that for this topology, a sequence $\{u_n\}$ in W converges if, and only if, $\{u_n\} \subset W_\alpha$ for some α and converges in this countably normed space.

PROBLEMS. (1) Show that the relative topology of X_1 as a subset of the topological space (X_2, τ_2) is a topology in the sense of

(1.2).

(2) Show that a sequence $\{u_n\}$ in a strict inductive limit of $\{W_\alpha\}$ which converges, actually converges in some W_α.

1.6 SPACES OF CONTINUOUS FUNCTIONS

A host of illuminating examples of normed and countably normed spaces is furnished by the continuous and continuously differentiable functions. These spaces also have important applications to the theory of differential equations where they are called test function spaces of one sort or another. We shall give an account of the principal ones in some detail in the next section. Leading up to this, we first prove a classical theorem.

DEFINITION 1.16. $C_o(R^n)$ will denote the set of continuous functions on R^n which tend to zero at infinity.

A precise way to characterize the idea of "tending to zero at infinity" is to state that for each $f \in C_o$, given an $\varepsilon > 0$, there exists a compact set $K(\varepsilon) \subset R^n$ such that $|f(x)| < \varepsilon$ whenever $x \in R^n - K$. A central result for these functions is

THEOREM 1.17. $C_o(R^n)$ is a Banach space with respect to the supremum norm

$$\| f \| = \sup_{x \in R^n} |f(x)| \quad . \tag{1.19}$$

<u>Proof</u>. As already mentioned in example 1.1 (d), the linear space

requirements are met by the usual rules for adding functions and multiplying them by complex numbers. It is also easy to check that (1.19) is actually a norm; therefore, we give only the proof of completeness. As this construction appears frequently in analysis, it is worth careful attention.

Consider a Cauchy sequence $\{f_n\}$ from C_o . For any $\epsilon > 0$, choose $n_o(\epsilon)$ such that

$$\sup_{x \in R^n} |f_n(x) - f_m(x)| < \epsilon \qquad m \geq n \geq n_o \ . \tag{1.20}$$

Thus at each point $x \in R^n$, $\{f_n(x)\}$ converges to a limit $f(x)$. It must be shown that $f \in C_o$. From (1.20)

$$\lim_{m\to\infty} \sup_{x \in R^n} |f_m(x) - f_n(x)| = \sup_{x \in R^n} |f(x) - f_n(x)| < \epsilon \tag{1.21}$$

and $\|f - f_n\| < \epsilon$ for n large enough. Hence, $\{f_n\}$ converges to f in the supremum norm. It is straight forward to show continuity of f . For pick $\epsilon > 0$ and consider

$$|f(x) - f(x')| \leq \sup_{x \in R^n} \{|f(x) - f_n(x)| + |f_n(x) - f_n(x')| + |f_n(x') - f(x')|\} \ .$$

Choosing n so that $\|f - f_n\| < \epsilon/3$ and $\delta(\epsilon)$ for which $|f_n(x) - f_n(x')| < \epsilon/3$ when $\|x - x'\| < \delta(\epsilon)$, we have

$$|f(x) - f(x')| < \epsilon \quad \text{if} \quad \|x - x'\| < \delta(\epsilon) \ .$$

Lastly, $f(x)$ tends to zero at infinity since $|f(x)| \leq |f(x) - f_n(x)| + |f_n(x)|$ and f_n tends to f uniformly on R^n .

Thus, the first term on the right may be made arbitrarily small uniformly in x by choice of n, while for x large enough the second term is small for any n. Hence $\lim_{x\to\infty} |f(x)| = 0$.

PROBLEMS. (1) Let $B(R^n)$ be the collection of bounded functions on R^n. Show that this is a Banach space with respect to the supremum norm.

(2) Let $B^\alpha(R^n)$ be the space of bounded functions defined on R^n satisfying a Lipschitz condition of order $0<\alpha<1$; namely, for any $x, x' \in R^n$ there exists a constant M such that $|f(x) - f(x')| \le M\|x - x'\|^\alpha$. Show that B^α is a Banach space with the norm

$$\| f \|_\alpha = \sup_{x,x'\in R^n} |f(x) - f(x')|/\|x - x'\|^\alpha + \sup_{x\in R^n} |f(x)| .$$

What is the relation between B, B^α, C_o, and $C(R^n)$?

(3) Show that $C^o([0,1])$ is a Banach space with the supremum norm. Generalize your result to $C(K)$, the continuous functions on any compact set K in R^n.

(4) Using the result of problem (3) show that $C^o([0,1])$ is separable.

Let us turn next to the continuously differentiable functions $C^m(R^n)$. In order to introduce a notion of convergence into this linear space, we should note that (1.19) or some of its generalizations involving derivatives, is not suitable. For example, polynomials lie in C^m, but do not have finite supremum over the whole of R^n. Equally, pointwise convergence on bounded sets will not do either. A sequence of differentiable functions converging pointwise need not have a differentiable limit, or even if it does, the limit may not be the limit of the derivatives. Examples of such pathology are easy to construct.

EXAMPLES. (a) Consider the sequence of functions defined by

$$f_n(x) = \begin{cases} 0 & x \le 0 \\ x \exp(\, -1/nx(1-x)\,) & 0 < x < 1 \\ 0 & x \ge 1 \,. \end{cases} \tag{1.22}$$

Each f_n is C^∞ due to the rapid decrease of the exponential at $x = 0,1$ which damps any polynomial growth at these points due to differentiation. However, as $n \to \infty$, the pointwise limit is the discontinuous function $f = x$ for $0 < x < 1$ and zero elsewhere. This is certainly not differentiable on R.

(b) To illustrate the second point, consider the sequence of functions $f_n(x)$ which are zero for $x \le 0$, x^n/n for $0 < x \le 1$, and $1/n$ for $x > 1$. These converge to zero for all x pointwise, but $f_n'(x)$ converges to a function which is 1 for $x = 1$ and zero elsewhere.

These two examples indicate that if we wish a sequence $\{f_n\} \subset C^m(R^n)$ to converge to a function which is again m-times continuously differentiable, and whose derivatives up to order m are limits of the sequences $\{D^r f_n\}$, $|r| \le m$ accordingly, then the convergence must be uniform on each bounded set of R^n, not only for f_n but also for each of the derivatives $D^r f_n$, $|r| \leq m$. The appropriate sense of convergence in C^m may then be spelled out in

DEFINITION 1.18. A sequence $\{f_n\}$ in $C^m(R^n)$ converges if the sequences $\{D^r f_n\}$, $|r| \leq m$ converge uniformly on every compact region $K \subset R^n$.

In this connection, a number of results are of interest concerning the convergence of derivatives and integrals.

LEMMA 1.19. Let $\{f_n\}$ be a sequence of continuous functions converging uniformly on the compact region K . Then

$$\lim_{n\to\infty} \int_K dx\, f_n(x) = \int_K dx \lim_{n\to\infty} f_n(x) .$$

Proof. From Theorem 1.17, $\lim_{n\to\infty} f_n(x) = g(x)$ exists and defines a continuous function on K . Consider the estimate

$$\left| \int_K dx\, g(x) - \int_K dx\, f_n(x) \right| \leq \|g - f_n\|\, m(K)$$

where $m(K)$ is the volume of the set K . As the right-hand side tends to zero for $n \to \infty$, the result follows.

LEMMA 1.20. Let $\{f_n\}$ converge to f in $C^m(R^n)$. Then $\{D^r f_n\}$ converges pointwise to $D^r f$ on every compact set K for all $|r| \leq m$.

Proof. By Definition 1.18 and Theorem 1.17, we know that

$\lim_{n\to\infty} D^r f_n = g_r$ exists for each $x \in K$ and all $|r| \le m$, and moreover, is continuous. For the case $r = 1$, Lemma 1.19 implies that $\int_a^x dt_i Df_n(t) \to \int_a^x dt_i g_1(t)$ if $D = \partial/\partial x_i$ as $n \to \infty$ for suitable x, a. But this is just

$$\lim_{n\to\infty} f_n(x) = f(x) = f(a) + \int_a^x dt_i \, g_1(t)$$

and hence by the Fundamental Theorem of Calculus, Df exists and is equal to g_1.

This type of argument may be repeated for each g_r whereupon $g_r = D^r f$ on all compacts K with $|r| \le m$.

Convergence in $C^m(R^n)$ as defined in 1.18 is interesting from the point of view of normed spaces. In fact, let us define

DEFINITION 1.21. $C^m(K)$ is the set of functions m-times continuously differentiable on the compact set $K \subset R^n$.

It is easy to see that $C^m(K)$ is normed by

$$\|f\|_{m,K} = \sup_{|r| \le m, x \in K} |D^r f(x)| \tag{1.23}$$

(see problem 1.1 (2)), and we might try to use these to cast definition 1.18 into the framework of countably normed spaces by the following device. Let

$$K_j = \{x \in R^n \mid \|x\| \le j, \; j = 1,2,3,\ldots\} \tag{1.24}$$

which is an increasing family of compact sets with $R^n = \bigcup_{j=1}^{\infty} K_j$.

Define a countable family by $\{\|\cdot\|_{m,K_j}\}$ and ask whether $C^m(R^n)$ is countably normed with respect to this family. The answer is clearly no, since (1.23) is not a norm on this space even though the family gives the correct sense of convergence. It is possible, however, to use these expressions to give a topology for $C^m(R^n)$ which leads to just the convergence in Definition 1.18 by introducing the concept of a semi-norm. This framework provides a more powerful and general set of techniques for studying linear topological spaces than those which we discuss. For them, we refer the reader to the literature cited at the end of the book, and to problems (7), (8) and (9) closing this section.

PROBLEMS. (5) Verify that $C^m(K)$ is normed by (1.23).

(6) Discuss the convergence in $C^m(R)$ of each of the following sequences of functions

(i) $x^{2m+1}\sin(1/nx)$ (ii) $\exp[(x^2 - 1/x^2)/n]$ (iii) $\log(1+x^2/n)$

(7) A function $p: V \to R_+$ is called a semi-norm for the linear space V if (i) $p(u+v) \le p(u)+p(v)$ (ii) $p(\alpha u) = |\alpha|\, p(u)$. Show that $p(0) = 0$ and $|p(u)-p(v)| \le p(u-v)$. A semi-norm contrasts with a norm in that $p(u) = 0$ need not imply $u = 0$.

(8) Show that $\|\cdot\|_{m,K_j}$ in (1.24) is a semi-norm for $C^m(R^n)$ (see (7)).

(9) Let p be a semi-norm for a linear space V (see problem 7). One knows the sets, $P_\varepsilon(u_o) = \{u \in V \mid p(u-u_o) < \varepsilon\}$, define a base for a topology for V. Show that the sense of con-

vergence in Definition 1.18 is correctly given by such a base with $\{\|\cdot\|_{m,K_j}\}$ as the collection of semi-norms.

1.7 SPACES OF TEST FUNCTIONS

Our last consideration in this chapter concerns spaces of C^∞ functions defined on R^n, the so-called testing functions. It is here that use will be made of the strict inductive limit studied in Definition 1.14 and Theorem 1.15. First, some general remarks and definitions will be needed.

DEFINITION 1.22. A function f defined on R^n is said to have support in a region S if $f(x) = 0$ whenever $x \in R^n - S$. The smallest such closed set S is called the support of f, written $\operatorname{supp} f$.

DEFINITION 1.23. Let $\mathcal{D}^m(K)$ denote the set of functions which have continuous derivatives up to order m, all having support in a compact region $K \subset R^n$.

It is clear that $\mathcal{D}^m(K)$ is a linear space with the usual operations (see example 1.1 (d)); in fact, $\mathcal{D}^m(K) \subset C^m(K)$ as a proper subset since the functions in the latter set need not have support contained in K. Some typical examples are given below.

EXAMPLES. (a) Suppose K is the segment $-1 \le x \le 1$. Then a function in $\mathcal{D}^m([-1,1])$, for all m, is given by

$$\theta_1(x) = \begin{cases} \exp(-1/1-x^2) & |x| \le 1 \\ 0 & |x| > 1 . \end{cases} \tag{1.25}$$

θ_1 is continuous at $|x| = 1$. This is also true for each of its derivatives which have the form $\theta_1^{(n)}(x) = P(x, 1/1-x^2)\theta_1(x)$ where $P(a,b)$ is a polynomial in a and b . For any x , $\exp|x| \ge |x|^N$ for any positive integer N ; thus, as $|x| \to 1$, we have $|\theta_1^{(n)}(x)| \le |P(x, 1/1-x^2)|(1-x^2)^N \to 0$ by choosing N sufficiently large for given n .

The function θ_1 is very useful in the theory of generalized functions in that it may be used to construct approximations to other continuous functions which need not have compact support.

(b) Functions in $\mathcal{D}^m([-1,1])$ for finite m are readily found by replacing the exponential in (1.25) by a power. For example, the function $f = (1-x^2)^{m+\frac{1}{2}}$ for $|x| \le 1$ and zero elsewhere.

For finite values of m , $\mathcal{D}^m(K)$ is a normed space with respect to (1.23); in fact, more is true, $\mathcal{D}^m(K)$ is a Banach space. Lemma 1.20 shows that a sequence $\{f_n\}$ converges to a function f which is m-times continuously differentiable as the norm topology agrees with the sense of convergence in Definition 1.18. The support of f must also be contained in K . For if not, there exists $x \notin K$ for which $f(x) \neq 0$. But by the uniform convergence of f_n to f , $|f(x) - f_n(x)| < \epsilon$ for any positive ϵ and sufficiently large n , a contradiction on the assumption $\operatorname{supp} f_n \subset K$. The same reasoning applies to the derivatives $D^r f$, $|r| \le m$.

For the applications which we have in mind for Chapter two, the space $\mathcal{D}^m(K)$ is of importance only as an intermediate step in the construction of a strict inductive limit of countably normed spaces. These are given in the next theorem.

THEOREM 1.24. $\mathcal{D}^\infty(K)$ is a complete countably normed space with respect to the family $\{\|\cdot\|_{m,K}\}_{m=1}^\infty$ of norms (1.23). Moreover, $\mathcal{D}^\infty(K) = \bigcap_{m=0}^\infty \mathcal{D}^m(K)$ is true.

Proof. The appropriate conditions on the norms are easy to check. The increasing strength follows by definition

$$\|f\|_{0,K} \le \|f\|_{1,K} \le \dots \le \|f\|_{m,K} \le \|f\|_{m+1,K} \le \dots$$

as well as their mutual compatibility. In fact, as a relation between sets $\mathcal{D}^\infty(K) = \bigcap_{m=0}^\infty \mathcal{D}^m(K)$ holds, whereupon Theorem 1.12 shows that $\mathcal{D}^\infty(K)$ is complete.

The first space of test functions to be examined is the set of C^∞ functions defined on R^n having compact support. More precisely

DEFINITION 1.25. Let $\mathcal{D}(R^n)$ denote the set of functions which belong to $\mathcal{D}^\infty(K)$ for some compact set $K \subset R^n$.

This is certainly a linear space in the customary way since for two such functions f_1, f_2 with $\operatorname{supp} f_1 \subset K_1$ and $\operatorname{supp} f_2 \subset K_2$ then $\operatorname{supp}(f_1 + f_2) \subset K_1 \cup K_2$ which is again compact if K_1 and K_2 have this property. This extends to finite sums $\Sigma_i\, \alpha_i\, f_i$ with complex coefficients. As we shall have occasion to refer to $\mathcal{D}(R^n)$ frequently, it is convenient to use the abbreviated notation $\mathcal{D}$ for this space. The main structure theorem for $\mathcal{D}$ is the content of our next result.

THEOREM 1.26. The space $\mathcal{D}$ is a strict inductive limit $\bigcup_{j=1}^{\infty} \mathcal{D}(K_j)$ where the compact sets K_j are given by (1.24).

Proof. In accordance with the definition of the strict inductive limit, $\mathcal{D}(K_j) \subset \mathcal{D}(K_{j+1})$ and the topologies of these spaces are compatible with such inclusions. For a sequence $\{f_n\}$ which converges to the zero function in $\mathcal{D}(K_j)$, necessarily converges to the same limit in $\mathcal{D}(K_{j+1})$ since $K_j \subset K_{j+1}$ and the support of the limit of the $\{f_n\}$ must lie in K_j. Thus, the inductive limit of the family $\{\mathcal{D}(K_j)\}$ is well defined and every function in $\mathcal{D}$ must belong to at least one of these spaces; and vice versa.

For most purposes the test functions $\mathcal{D}$ will be the most restrictive class that we shall need to consider. Numerous other test functions result by modifying the support and growth properties of C^∞ functions. A further example will be given below, but first some remarks to amplify the manner in which sequences of functions converge in $\mathcal{D}$. Our comments following Theorem 1.15 indicate that a sequence $\{f_n\} \subset \mathcal{D}$ converges to a limit f, again in $\mathcal{D}$, if the convergence takes place in some $\mathcal{D}(K_j)$ for j sufficiently large. From Theorem 1.24 and the form of the norm (1.23), this implies that

(a) there exists a compact set K such that $\operatorname{supp} f_n \subset K$ for all n

(b) for each multi-index r , $D^r f_n \to D^r f$ uniformly in K .

The strict inductive limit is a piecing together of the spaces $\mathcal{D}(K_j)$ in such a way that the topologies remain consistent on the sets $\mathcal{D}(K_j) \cap \mathcal{D}(K_i)$.

PROBLEMS. (1) Show that the function, $f(x) = \exp(-1/x^2) \cdot \exp(-1/(x-a)^2)$ for $0 \leq x \leq a$ and zero elsewhere, is in $\mathcal{D}(R)$. Generalize this to $\mathcal{D}(R^n)$.

(2) Consider the function $\theta(x) = c\theta_1(x/a)$ for $a > 0$. Show that it is in $\mathcal{D}$ and find its support. Determine the constant c so that $\int_{R^n} dx\ \theta(x) = 1$.

(3) Check convergence in $\mathcal{D}$ for the sequence $f_n(x) = \theta_1(x/a)/n$ and find the limit. Does $g_n(x) = \theta_1(x/na)/n$ also converge in $\mathcal{D}$?

(4) Verify the compatibility of the norms (1.23) for $\mathcal{D}(K)$.

(5) A function f defined on R^n is said to be real analytic at a point x_o if there exists a power series $f(x) = \sum_{n=0} a_n(x - x_o)^n$ which is uniformly convergent in some neighborhood of x_o . Show that $\mathcal{D}$ does not contain any real analytic functions. (Hint: Take the case of $n = 1$ first and use the power series in a neighborhood of x_o to define a function $f(x) = f(x+iy)$ of a complex variable z in a complex neighborhood of x_o . Then use the principle of analytic continuation to show that the only real analytic function with compact support is the zero function.)

Our final discussion of the chapter concerns one further class of test functions: the C^∞ functions of rapid decrease. These arise in a natural way in Fourier analysis and its applications, and as such, they will appear throughout much of our later work. First, we define what is meant by rapid decrease.

DEFINITION 1.27. A function f defined on R^n is said to be of rapid decrease if

$$\lim_{\|x\|\to\infty} \|x\|^N \, |f(x)| = 0$$

for all positive integers N. The norm $\|x\|$ is the Eculidean distance.

Particular cases of such functions are easy to write down; for example, on R $\exp(-x^2)$, $1/\cosh x$, or any function of compact support. On R^n these generalize to $\exp(-\|x\|^2)$, $\prod_{i=1}^{n} 1/\cosh x_i$. From the definition, a function of rapid decrease may have arbitrary support, but at large distance it must tend to zero more rapidly than any polynomial. It is immediate that for a polynomial $P(x)$, $\lim_{\|x\|\to\infty} |P(x)\, f(x)| = 0$ when f has rapid decrease.

The collection of continuous functions having rapid decrease forms a linear space which we denote by $\mathcal{S}^0(R^n)$. The topological structure is revealed by means of the norms

$$\|f\|^s = \sup_{x\in R^n,\ |q|\le s} |x^q f(x)| \qquad s = 0,1,2,\ldots \tag{1.26}$$

in the following way.

DEFINITION 1.28. Let $\mathcal{S}_s^0(R^n)$ denote the continuous functions defined on R^n and normed with respect to (1.26).

Then $\mathcal{S}_s^0$ is a Banach space for each positive integer s and $\mathcal{S}^0 = \bigcap_{s=0}^{\infty} \mathcal{S}_s^0$, whereupon we learn that $\mathcal{S}^0$ is a complete countably normed space. The proofs of these statements are left as an exercise. Spaces of differentiable functions having fast decrease are similarly defined by extending the above definition in an obvious manner.

DEFINITION 1.29. Let $\mathcal{S}^m(R^n)$ denote the space of functions m-times continuously differentiable, which together with their derivatives, are of rapid decrease.

As the reader can see, the spaces of differentiable functions having various decrease properties can be built up in equivalent ways. Either one may impose smoothness requirements on $\mathcal{S}^0$ or one may require rapid decrease of functions in the class C^m. It is then easy to see how the various compatible norms are to be chosen. If we define

$$\| f \|_m^s = \sup_{x \in R^n,\ |r| \leq m, |q| \leq s} |x^q D^r f(x)| \tag{1.27}$$

then we have the result

THEOREM 1.30. $\mathcal{S}^m(R^n)$ is a complete countably normed space expressed as $\mathcal{S}^m(R^n) = \bigcap_{s=0}^{\infty} \mathcal{S}_m^s(R^n)$, where $\mathcal{S}_m^s(R^n)$ is a Banach space of differentiable functions, normed by (1.27).

With these preliminaries, the space of test functions of rapid decrease may be given as a countably normed space.

DEFINITION 1.31. The space $\mathcal{S}(R^n)$ of C^∞ functions of rapid decrease is $\bigcap_{m=0}^{\infty} \mathcal{S}^m(R^n)$.

For historical reasons, this space is often referred to as the "Schwartz" space in honor of the French mathematician, Laurent Schwartz, who first introduced a systematic theory of generalized functions and test function spaces.

From the definitions given above, it is clear that $\mathcal{D} \subset \mathcal{S} \subset C^\infty$ with these inclusions valid for the corresponding topologies. Namely, a sequence $\{f_n\}$ which converges in the sense of $\mathcal{D}$ also converges in the sense of $\mathcal{S}$. A sequence converging in $\mathcal{S}$ also converges in C^∞ . These are easy to verify.

PROBLEMS. (6) Show that $\mathcal{S}(R)$ is separable. Can you extend this to R^n ? (Hint: First show that each $\mathcal{S}^m(R^n)$ is separable.)

(7) In the inclusions $\mathcal{D} \subset \mathcal{S} \subset C^\infty$, show that converges in the smaller space implies convergence in the larger space.

(8) Show that the function $\exp(i\|x\|^2)$ lies in C^∞ , but neither in $\mathcal{D}$ nor in $\mathcal{S}$.

The notion of completeness introduced for a normed space in Definition 1.7 extends to any topological space. The space (X, τ)

is said to be complete if any Cauchy sequence has a limit with respect to the topology τ . For the test function spaces, completeness is a central result.

THEOREM 1.32. Each of the test function spaces $\mathcal{D}$, $\mathcal{S}$, or C^∞ is complete with respect to their topologies.

Proof. For $\mathcal{S}$, completeness is a consequence of Theorems 1.12 and 1.30.

The case of $\mathcal{D}$ is readily handled. Let $\{f_n\} \subset \mathcal{D}$ be a Cauchy sequence, then $\{f_n\}$ is Cauchy in some $\mathcal{D}(K_j)$ which is complete by Theorem 1.24. Thus, $f_n \to f$ as elements of $\mathcal{D}(K_j)$; hence, also in $\mathcal{D}$ from the properties of strict inductive limits.

Finally, suppose $\{f_n\}$ is Cauchy in C^∞ . Then $\{f_n\}$ is Cauchy in each $C^m(K)$, where K is any compact subset of R^n . But $C^m(K)$ is complete by Lemma 1.20. Thus, for each compact K , there exists a function f_K such that $D^r f_n \to D^r f_K$ uniformly on K for all multi-indices r . Consider two such compact sets K_1 and K_2 with non-empty intersection. Then for $x \in K_1 \cap K_2$, we have the estimate

$$|D^r f_{K_1}(x) - D^r f_{K_2}(x)| \le |D^r f_{K_1}(x) - D^r f_n(x)| + |D^r f_n(x) - D^r f_{K_2}(x)|$$

which may be made arbitrarily small by choosing n large enough. The functions $\{f_K\}$ may then be pieced together in a consistent way to define a function $f \in C^\infty(R^n)$. Clearly, this f is the limit of $\{f_n\}$ in the sense of Definition 1.18 for any m .

PROBLEM (9) Give an explicit proof of the completeness for $\mathcal{S}$ which uses the definition of $\mathcal{S}$ as C^{∞} functions with rapid decrease and not Theorem 1.12. Namely, $\{f_n\}$ is Cauchy if

$$\lim_{p,q\to\infty} \sup_{x} |x^s D^r(f_p - f_q)| = 0 \text{ for all } r \text{ and } s .$$

2. GENERALIZED FUNCTIONS

Generalized functions are defined as continuous linear functionals on the spaces of test functions $\mathcal{D}$, $\mathcal{S}$, C^∞ . The rules for differentiation and convergence follow easily and lead to the result that generalized functions are limits in some suitable topology of functions in $\mathcal{D}$. Several properties, such as convolution and completeness, are also discussed.

2.1 THE NOTION OF A GENERALIZED FUNCTION

In many applications of analysis to mathematical problems in physics or engineering, it is necessary to deal with discontinuous or singular functions, such as step functions, δ-pulses, and their derivatives. It is as well to have a systematic mathematical framework for such operations with precise control over them and the relations between them. Before undertaking this development, let us examine some simple examples.

For the function e^x , the improper integral

$$\lim_{R\to\infty} \int_{-R}^{R} dx\ e^x$$

does not exist due to the increase of the exponential as $x \to \infty$. This rather dramatic divergence can be controlled if we consider the integral

$$\langle e^x , f(x) \rangle = \int_{-\infty}^{\infty} dx\ e^x\ f(x) = \int_a^b dx\ e^x\ f(x) \qquad (2.1)$$

where $f \in \mathcal{D}(R)$ with $\operatorname{supp} f \subset [a,b]$; $-\infty < a < b < \infty$. In (2.1) we have chosen to give up the use of e^x pointwise, regarding it rather as defining a linear functional on $\mathcal{D}$. Notice also, that if $\{f_n\}$ is a sequence tending to zero in $\mathcal{D}$, there exist a and b as above such that $\operatorname{supp} f_n \subset [a,b]$ uniformly in n and $\langle e^x , f_n(x) \rangle \to 0$ as $n \to \infty$. Thus, e^x considered as a linear functional has a certain continuity property with respect to the test functions over which it is defined.

Whether or not (2.1) for any $f \in \mathcal{D}(R)$ provides a satisfactory definition of the integral of e^x over the whole real line depends upon the applications which we have in mind and whether the "weak" knowledge $\langle e^x , f(x) \rangle$ is adequate. As we shall see, it is possible to reconstruct the function e^x pointwise on any bounded set from expressions of the form (2.1); moreover, a certain uniqueness holds in the sense that if $\langle e^x - \phi(x) , f(x) \rangle = 0$ for all $f \in \mathcal{D}(R)$, then ϕ and e^x are equal in some sense.

A second example indicates how one might deal with integrals involving functions which are singular for finite values of their arguments. Consider the function $1/x^2$ and its improper integral

$$\lim_{\delta \to 0^+} \int_{-1}^{-\delta} + \int_{\delta}^{1} dx\ 1/x^2$$

which does not exist. However, suppose $f \in C^\infty$, then

$$\lim_{\delta \to 0^+} \int_{-1}^{-\delta} + \int_{\delta}^{1} dx[f(x) - f(0) - x\, f'(0)]/x^2 \tag{2.2}$$

has a well-defined limit using Taylor's theorem on $f(x)$ near the origin. A generalization to $1/x^n$ might lead one to consider the expression

$$\lim_{\delta \to 0^+} \int_{-1}^{-\delta} + \int_{\delta}^{1} dx\, \frac{f(x) - f(0) - xf'(0) - \ldots - x^{n-1} f^{(n-1)}(0)/(n-1)!}{x^n} \tag{2.3}$$

which also exists. In both (2.2) and (2.3) an essential role is played by the differentiability or smoothness of the test functions near the point of singularity, while in (2.1) the compact support or rate of decrease at infinity was the determining factor.

In spite of the contrived nature of these two examples, they do indicate that a natural interpretation may be given to ordinary functions which lack continuity, or integrability, for certain ranges of their arguments by regarding them as linear functionals on test functions. If (2.3) is to work for arbitrary positive integers n, then f should be C^∞ near zero. Similarly $f \in \mathcal{D}$ in (2.1) would permit us to integrate any smooth function with arbitrary rate of increase at infinity; for example, $\exp(\exp(\exp x))$. These two regularity requirements would allow us to define a wide class of linear functionals; in fact, too wide a class for most applications. It is essential that these functionals should have the weak continuity property illustrated above in connection with (2.1). By reference to definition 1.18 the reader will have no difficulty in verifying a similar property for (2.3).

As a last illustration we turn to a physical example: the measurement of a static electric field $\vec{E}(\vec{r})$ in a region K of three dimensional space. For purposes of discussion, it is sufficient to take K bounded. In electrodynamics $\vec{E}$ is defined as the limit

$$\vec{E}(\vec{r}_o) = \lim_{\delta q \to 0} \vec{F}(\vec{r}_o)/\delta q \tag{2.4}$$

where $\vec{F}(\vec{r}_o)$ is the force on a test charge of magnitude δq placed at the point $\vec{r}_o \in K$. This expression is an idealization in the sense that any test probe carrying charge δq can never be localized precisely at a point experimentally, but must have a finite extension in space.

An attempt at a more precise mathematical characterization of the electric field along the lines of (2.4) might begin as follows. Let $f(\vec{r})$ be a test function from $\mathcal{D}$ such that $\vec{r}_o \in \operatorname{supp} f \subset K$ and $\int d\vec{r}\, f(\vec{r}) = 1$; for example, $\theta(\vec{r} - \vec{r}_o)$ in problem 1.7 (2). A charge distribution localized at $\vec{r}_o$ with total charge δq is then described by $\delta q\, f(\vec{r})$. Experimentally the average force per unit charge on this distribution is a linear functional which we take to be the average electric field at $\vec{r}_o$ and write

$$\langle \vec{E}(\vec{r}_o) \rangle_{avg} = \langle \vec{E}(\vec{r}), f(\vec{r}) \rangle = \int d\vec{r}\, \vec{E}(\vec{r})\, f(\vec{r}) \ . \tag{2.5}$$

As the charge distribution, which is used to measure the electric field, is arbitrary in shape, we must allow all test functions $f \in \mathcal{D}(R^3)$ to appear in (2.5). We might attempt to recover the pointwise expression (2.4) by taking the limit

$$\lim_{a \to 0^+} \langle \vec{E}(\vec{r}) , \theta(\vec{r} - \vec{r}_o) \rangle \tag{2.6}$$

in which a, as introduced in problem 1.7 (2), is the "size" of the charge probe. When this limit exists, it could serve as a definition of $\vec{E}(\vec{r}_o)$. Clearly the existence of (2.6) is a stronger condition on the regularity of the electric field $\vec{E}$ than just the existence of the functionals (2.5), but accords with our intuition that the electric field should be independent of the probe which is used to measure it. In fact, one finds in the classical theory of electromagnetism that for most charge distributions which give rise to electric fields in the region K, the pointwise limit (2.6) does exist except at isolated points and reduces to the usual expressions. However, in the quantum electrodynamics of point particles, this is not always the case and one must adhere to expressions of the form (2.5). In this latter case, the nature of local interactions between point particles described by quantum mechanics requires singular expressions for the electromagnetic fields. These have to be understood within the framework of the theory of generalized functions.

We turn next to the systematic development of the mathematical ideas which will enable us to deal with expressions such as (2.1), (2.3) and (2.5).

2.2 GENERALIZED FUNCTIONS

In the following $\mathcal{J}$ will denote any one of the spaces of test functions introduced in section 1.7. For much of our formal work, it is immaterial which of these spaces is used. Exceptions will be indicated as they occur.

DEFINITION 2.1. A linear functional ϕ on $\mathfrak{J}$ will be said to be continuous, if for any sequence $\{f_n\}$ of test functions from $\mathfrak{J}$ converging to zero in the topology for $\mathfrak{J}$, the sequence of complex numbers $\phi(f_n) \to 0$ as $n \to \infty$. The set of continuous linear functionals on $\mathfrak{J}$ will be denoted by $\mathfrak{J}'$.

It has already been noted that $\mathfrak{J}'$, being a subspace of $L(\mathfrak{J}, \mathbb{C})$, is a linear space. Strictly speaking, the above definition is incomplete without the notion of convergence to be introduced by Definition 2.6. We have thought is best, however, to make complete definitions at this point and introduce new concepts leisurely. Topological considerations for the continuous linear functionals are to be discussed more fully in section 2.5.

DEFINITION 2.2. The elements of $\mathfrak{J}'$ will be called generalized functions for the test functions $\mathfrak{J}$.

The cases which we shall consider possess a particular terminology: the space $\mathfrak{D}'$ is called the space of distributions while the elements of $\mathfrak{S}'$ are called tempered distributions. Finally, $C^{\infty}{}'$ is referred to as the space of distributions with compact support. The reasons for this nomenclature will become clearer later on. By way of notation, we shall write $\phi(f)$ or $\langle \phi, f \rangle$ for the value of the functional ϕ on the test function f.

It is perhaps as well at this stage of our discussion to emphasize that the generalized functions defined above only take a meaning when applied to test functions. They are not always defined pointwise. The examples of section 2.1 indicate that the functions of ordinary analysis appear as a subclass among the generalized functions if we understand that for these

$$\langle \phi , f \rangle = \int dx \ \phi(x) \ f(x) \ . \tag{2.7}$$

This notation is so useful that it will frequently be abused and applied when ϕ is not defined pointwise. Now to further developments.

DEFINITION 2.3. Two generalized functions $\phi_1, \phi_2 \in \mathfrak{J}'$ are said to be equal if $\phi_1(f) = \phi_2(f)$ for all $f \in \mathfrak{J}$.

Though Definition 2.2 allows us to consider a wide class of quantities for which $\phi(f)$ exists, we can easily see that in particular cases Definition 2.3 shows that generalized functions may agree with ordinary functions. For example, suppose that g is integrable over any finite interval. Then

$$\langle g , f \rangle = \int dx \ g(x) \ f(x) \qquad f \in \mathfrak{D}(R) \tag{2.8}$$

defines g as a distribution. In general we shall not make explicit this distinction as it is readily understood within the context that it appears.

EXAMPLES. (a) One of the most important generalized functions which does not agree with an ordinary function is the Dirac δ-distribution defined by

$$\delta_x(f) = \langle \delta_x , f \rangle = f(0) \qquad f \in \mathcal{D} \tag{2.9}$$

and denoted δ_x . This generalized function picks out the value of a testing function at the origin, and as such, is defined not only for any of the spaces $\mathfrak{J}$ but also on any function which is continuous in some neighborhood of the origin. To investigate δ_{ax+b} in one dimension, choose $f \in \mathcal{D}$ and form

$$\langle \delta_{ax+b} , f \rangle = \int dx \ \delta(ax+b) \ f(x) = \int dy \ \delta(y) f((y-b)/a)/|a| \quad a \neq 0$$

$$= f(-b/a)/|a| \ .$$

Thus, in the sense of generalized functions, we define

$$\delta_{ax+b} = \delta_{x+b/a}/|a| \qquad a \neq 0 \tag{2.10}$$

with the consequence

$$\delta_x = \delta_{-x} \qquad (\delta \text{ is an even distribution}) \ . \tag{2.11}$$

Of particular interest is the case when $a = 1$, $b = x_o$; namely,

$$\langle \delta_{x-x_o} , f \rangle = f(x_o)$$

or $\delta(x - x_o)$ "shifts" the value of f to that at x_o .

PROBLEMS. (1) Verify $\delta(x^2 - a^2) = [\delta(x - a) + \delta(x + a)] / 2|a|$ with $a \neq 0$ holds in the sense of Definition 2.3.

(2) Let $\alpha(x)$ vanish at a finite number of points

$x_k \in R$ with $0 < |\alpha'(x_k)| < \infty$ for each k. Show that $\delta_{\alpha(x)} = \sum_k \delta(x - x_k) / |\alpha'(x_k)|$.

EXAMPLE. (b) The δ-distribution is extended to R^n by taking x in (2.9) an n-dimensional vector and interpreting

$$\delta_{(ax+b)} = \prod_{i=1}^{n} \delta_{ax_i+b_i} \tag{2.12}$$

as a product in the n-variables separately. Results analogous to (2.10) and (2.11) hold; however, if $\alpha(x) = (\alpha_1(x), \ldots, \alpha_n(x))$, then

$$\delta_{\alpha(x)} = \sum_k \delta_{x-x_k} / |J(\alpha| x_k)|$$

where $J(\alpha| x) = \det[\partial \alpha_i(x) / \partial x_j]$ is the Jacobian of $\alpha(x)$ and $x_k = (x_{k1}, \ldots, x_{kn})$ are the zeroes of $\alpha(x)$. Clearly we assume $J(\alpha| x_k) \neq 0$.

In the two examples above, we have used test functions in $\mathcal{D}$; we could have equally well taken test functions from either $\mathcal{S}$ or C^∞. The δ-distribution is an example of a generalized function having discrete support since it yields only those values of the appropriate test functions at a single point. This observation is useful for defining the support of a generalized function.

DEFINITION 2.4. A $\phi \in \mathcal{J}'$ is said to have support in S if $\phi(f) = 0$ whenever $f \in \mathcal{J}$ has support outside of S. The smallest closed set with this property is called the support of ϕ, written supp ϕ . When supp ϕ is compact, ϕ is said to be a generalized function of compact support.

It may happen that supp ϕ consists of a finite, or countably infinite, set of points in R^n . In this case we shall say that ϕ has discrete support and indicate whether it is finite or infinite. Clearly δ_x has discrete support at the origin, while as we shall see, $\Sigma_{n=1}^{\infty} \delta(x-1/n)$ has as support the set $\{1,1/2,\ldots,1/n,\ldots\}$. Other examples using the δ-distribution are readily constructed.

As a last remark in this section, suppose $u(x) \in C^{\infty}$ and $\phi \in \mathcal{D}'$. Then we may multiply the distribution ϕ by $u(x)$ to obtain a new element $u\phi$ of $\mathcal{D}'$ defined by

$$\langle u\phi , f \rangle = \langle \phi , u\, f \rangle \qquad f \in \mathcal{D} . \tag{2.13}$$

This is a consistent definition as uf is again a test function from $\mathcal{D}$ due to the continuous differentiability of u to all orders. Moreover, if $\{f_n\}$ is a sequence tending to zero in $\mathcal{D}$, so is $\{uf_n\}$, and (2.13) implies that $\phi(uf_n) = u\phi(f_n) \to 0$. Thus, $u\phi$ has the requisite continuity to be a member of $\mathcal{D}'$. Generalizing (2.13) to arbitrary test function spaces leads to the notion of multiplier space.

DEFINITION 2.5. Let $\mathcal{J}$ be a space of test functions. The multiplier space of $\mathcal{J}$ is the set of functions $\mathfrak{M}(\mathcal{J}) = \{u \in C^{\infty} | uf \in \mathcal{J} \text{ for all } f \in \mathcal{J}\}$.

The property characteristic of the multiplier space is that for any

ϕ , $u\phi$ is again in $\mathfrak{J}$ for all $u \in \mathfrak{m}(\mathfrak{J})$ by means of (2.13). This is not quite as obvious as it might seem at first sight. It must also be checked that $u\phi$ is continuous on $\mathfrak{J}$ in the sense of Definition 2.1. This is given as an exercise below.

PROBLEMS (3) Show that $x\delta_x = 0$; hence, if $P(x)$ is any polynomial in x with constant term a_o ; $P(x)\delta_x = a_o\,\delta_x$. What is the support of $P(x)\delta_x$?

(4) What is the support of $\phi = \delta(\sin x)$? Is ϕ in all of the spaces $C^{\infty\prime}$, $\mathfrak{S}'$, $\mathfrak{D}'$?

(5) Prove that $\mathfrak{m}(\mathfrak{D}) = C^{\infty} = \mathfrak{m}(C^{\infty})$, while $\mathfrak{m}(\mathfrak{S}) = \{u \in C^{\infty} | \, |u(x)/(1+\|x\|)^N|$ is bounded for some integer N , $x \in R^n\}$.

2.3 CONVERGENCE OF GENERALIZED FUNCTIONS

As before, let $\mathfrak{J}$ denote any of the test function spaces under consideration and $\{\phi_n\}$ a sequence of generalized functions from $\mathfrak{J}'$.

DEFINITION 2.6. The sequence $\{\phi_n\}$ converges to $\phi \in \mathfrak{J}'$, written $\phi_n \to \phi$, if the sequence of complex numbers $\langle \phi_n - \phi , f \rangle$ converges to zero for all $f \in \mathfrak{J}$.

It is easy to see that the usual results hold; namely, if $\phi_n \to \phi$ and $\psi_n \to \psi$ are convergent in $\mathfrak{J}'$, then also $\phi_n + \alpha\psi_n \to \phi + \alpha\psi$ for any $\alpha \in \mathbb{C}$ and $u\phi_n \to u\phi$ with $u \in \mathfrak{m}(\mathfrak{J})$ are convergent in $\mathfrak{J}'$.

Given a series of generalized functions, say $\Sigma_{n=1}^{\infty} \phi_n$, we shall say that this converges if the sequence formed from the partial sums converges. We write $\phi = \Sigma_{n=1}^{\infty} \phi_n$ if $\lim_{N\to\infty} \langle \Sigma_{n=1}^{N} \phi_n , f\rangle = \phi(f)$ for all $f \in \mathcal{J}$. The rules for adding convergent series and multiplying them term by term with functions in the multiplier space follow from the like rules for sequences stated above.

EXAMPLE A simple example of convergence for generalized functions is the series of δ-distributions $\Sigma_{n=1}^{\infty} \delta(x-n)$. For a test function f, we must examine the convergence of the series

$$\langle \sum_{n=1}^{\infty} \delta(x-n) , f(x)\rangle = \sum_{n=1}^{\infty} f(n) . \tag{2.14}$$

When $f \in \mathcal{D}$, the compact support property implies that for some $n_o(f)$, $f(n) = 0$ whenever $n > n_o$. Hence, there are only a finite number of terms in the series on the right-hand side of (2.14). If $f \in \mathcal{S}$, the rapid decrease guarantees convergence. Thus $\Sigma_{n=1}^{\infty} \delta(x-n)$ converges in either $\mathcal{D}'$ or $\mathcal{S}'$. However, this is no longer true for $C^{\infty\prime}$. For example, x^2 or e^x are perfectly good members of C^{∞}, but neither of the series $\Sigma_{n=1}^{\infty} n^2$, $\Sigma_{n=1}^{\infty} e^n$, which appear in (2.14), converge. Even though each member $\delta(x-n)$ lies in $C^{\infty\prime}$, the sum over n does not. This example also serves to illustrate an important point concerning the support of generalized functions in $C^{\infty\prime}$. The difficulty with $\Sigma_{n=1}^{\infty} \delta(x-n)$ on C^{∞} is due to the unboundedness of its support which is discrete, just the set $\{1,2,3,\ldots\}$. It will be a general feature of $\phi \in C^{\infty\prime}$ that $\operatorname{supp} \phi$ must be a compact set in R^n.

In section 1.4 we introduced the important notion of completeness of a normed space with respect to Cauchy sequences. The same questions arise for spaces of generalized functions and lead us to define the Cauchy property for sequences in $\mathfrak{J}'$.

DEFINITION 2.7. A sequence $\{\phi_n\} \subset \mathfrak{J}'$ is said to be Cauchy if the sequences $\{\phi_n(f)\}$ of complex numbers are Cauchy for any $f \in \mathfrak{J}$.

Clearly if $\phi_n \to \phi$ in the sense appropriate to $\mathfrak{J}'$, then $\{\phi_n\}$ is Cauchy. Most important is the fact that the converse is also true; namely, $\mathfrak{J}'$ is complete. Proof of this fact will be given in section 2.5.

PROBLEMS (1) Study the convergence of the series $\Sigma_{n=0}^{\infty}\ \delta(x-(2n+1)\pi/2)$, $\Sigma_{n=0}^{\infty}\ e^x\ \delta(x-n)$ on the spaces $\mathfrak{J}$.

(2) Study the convergence in $\mathfrak{J}'$ of each of the sequences $\delta(e^{nx})$, $\delta(e^{x/n})$, $\delta(\sin(x/n)/n)$ for $n = 1,2,3,\ldots$.

2.4 DERIVATIVES OF GENERALIZED FUNCTIONS

Having dispensed with some of the basic definitions and elementary properties of generalized functions, we should like to define an extended notion of derivative. Since ordinary functions may be special cases of generalized functions, we should require that this extended notion of derivative agree with the customary one where possible. In the following, multi-index notation continues to be used with the symbol D (see section 1.1) for derivative.

DEFINITION 2.8. For $\phi \in \mathfrak{J}'$, the derivative $D\phi$ is the generalized function given by $D\phi(f) = -\phi(Df)$ for all $f \in \mathfrak{J}$.

As test functions are C^∞, higher order derivatives exist and satisfy

$$\langle D^r \phi , f \rangle = (-1)^{|r|} \langle \phi , D^r f \rangle \text{ all } f \in \mathfrak{J} . \tag{2.15}$$

One may verify the linearity and continuity of $D^r\phi$ as defined by (2.15) with the result that $\phi \in \mathfrak{J}'$ implies $D^r\phi$ is also in $\mathfrak{J}'$. As there is no restriction on the magnitude of the indices r_i appearing in (2.15), we see that Definition (2.8) implies that differentiation of generalized functions is possible to any order. This is clearly a consequence of requiring test functions to be C^∞ with topologies which preserve this smoothness (see Definition 1.18).

For an ordinary function $g(x)$ which together with its derivative is integrable over any finite interval, we define a distribution by means of (2.8). Then upon integrating by parts

$$\langle Dg , f \rangle = -\langle g , Df \rangle = -\int dx \; g(x) \; f'(x) = \int dx \; g'(x) \; f(x) \; .$$

Thus Dg and dg/dx agree wherever the latter exists. Definition 2.8 then satisfies our requirement of consistancy.

EXAMPLES (a) <u>The Heaviside Step Function $H(x)$</u>.

Consider the function $H(x)$ defined by

$$H(x) = \begin{cases} 1 & x > 0 \\ 0 & x < 0 \end{cases} \qquad (2.16)$$

Regarding $H(x)$ as a generalized function, our remark above gives $DH = 0$ for $x \neq 0$. In fact, $DH(x) = \delta_x$ since for any $f \in \mathcal{D}$, $DH(f) = -H(Df) = -\int_0^\infty dx\, f'(x) = f(0) = \delta_x(f)$.

Corresponding step functions in many variables may be defined by multiplying together step functions in each of the variables separately. For example, in R^n, $H(x) = H(x_1)\, h(x_2) \ldots H(x_n)$ has support in the sector $\{x \in R^n \mid x_i > 0 ,\ 1 \le i \le n\}$ and value one for points in this set. A simple extension of the previous argument leads to $D^{(1,1,\ldots,1)} H(x) = \delta_x$ in n-variables.

For steps at points $x_o \neq 0$, choose $H(x - x_o)$ in the case of one variable. The extension to R^n is just a product of similar expressions.

(b) <u>The sign function sgn x</u>.

The sign function in one variable is defined by

$$\operatorname{sgn} x = \begin{cases} 1 & x > 0 \\ -1 & x < 0 \end{cases} \qquad (2.17)$$

An alternative expression is $\operatorname{sgn} x = H(x) - H(-x)$, whereupon it follows that $D \operatorname{sgn} x = 2\delta_x$.

(c) <u>$D^r\, \delta_x$</u>.

For any $f \in \mathfrak{J}$, we may evaluate $D^r\, \delta_x$ by means of (2.15); namely,

$$\langle D^r \delta_x, f\rangle = (-1)^{|r|}\langle \delta_x , D^r f\rangle$$
$$= (-1)^{|r|}\partial^{|r|} f(x)/\partial_{x_1}^{r_1}\partial_{x_2}^{r_2}\ldots\partial_{x_n}^{r_n}\Big|_{x=0} \quad . \qquad (2.18)$$

PROBLEMS (1) Show that $x^m D^r \delta_x = 0$ if $|m| \geq |r| + 1$.

(2) Let $u \in \mathfrak{m}(\mathfrak{J})$. Prove that for $\phi \in \mathfrak{J}'$, $D(u\phi) = u(D\phi) + (Du)\phi$ as a relation between generalized functions.

(3) Verify Liebnitz's formula for generalized functions $D^r(u\phi) = \Sigma_{k=0}^{r} \binom{r}{k} (D^k\phi)(D^{r-k}u)$ where $u \in \mathfrak{m}[\mathfrak{J}(R)]$. Extend this to generalized functions in $\mathfrak{J}'(R^2)$.

As a further example of the relation between the ordinary derivative and the derivative in the sense of generalized functions, we prove a useful result in the case of one variable.

PROPOSITION 2.9. Let $g(x)$ be a piecewise continuous function defined on R . Suppose that g has a finite number of discontinuities at the points $a_1, a_2, \ldots, a_k$ of magnitude $\alpha_1, \alpha_2, \ldots, \alpha_k$, and that dg/dx is piecewise continuous between the jumps. Then if $g \in \mathfrak{J}'$

$$Dg = dg/dx + \sum_{p=1}^{k} \alpha_p \, \delta_{x-a_p} \quad . \qquad (2.19)$$

Proof. Consider the function $h(x) = g(x) - \Sigma_{p=1}^{k} \delta\alpha_p H(x - a_p)$. It is continuous and its ordinary derivative, where it exists, agrees with the generalized function derivative. Hence, $Dh = h'(x) = g'(x) = Dg - \Sigma_{p=1}^{k} \alpha_p \delta(x - a_p)$, which is just (2.19).

COROLLARY. If $b_1, b_2, \ldots, b_m$ are points of discontinuity in $g'(x)$ with jumps of magnitude $\beta_1, \beta_2, \ldots, \beta_m$, then

$$D^2 g = d^2 g/dx^2 + \sum_{p=1}^{k} \alpha_p \, \delta'(x - a_p) + \sum_{q=1}^{m} \beta_q \, \delta(x - b_q) \quad . \tag{2.20}$$

EXAMPLE (d) Let $g(x) = \cos x$ for $x > 0$ and zero elsewhere. Then by (2.19), $Dg = -H(x) \sin x + \delta(x)$.

PROBLEM (4) Consider the function $|x| = x$ for $x > 0$ and $-x$ for $x < 0$. Show that $D|x| = \operatorname{sgn} x$, $D^2 |x| = 2\delta(x)$.

For applications to problems involving the solution of differential equations by means of generalized functions expressed as infinite series, it is central that convergence be sufficient to imply that a series differentiated term by term in the sense of generalized functions, also converges in one of the spaces $\mathfrak{J}'$. This property lies at the heart of Fourier series methods applied to such problems (see Chapter three). In this respect, we prove that series of generalized functions are very well-behaved.

LEMMA 2.10. Consider a sequence $\{\phi_n\}$ which converges to $\phi \in \mathfrak{J}'$. Then $\{D^r \phi_n\}$ converges in $\mathfrak{J}'$ to $D^r \phi$ for all multi-indices r .

<u>Proof</u>. For any test function $f \in \mathfrak{J}$, we have

$$\langle D^r \phi_n , f \rangle = (-1)^{|r|} \langle \phi_n , D^r f \rangle \xrightarrow[n \to \infty]{} (-1)^{|r|} \langle \phi , D^r f \rangle = \langle D^r \phi , f \rangle \quad .$$

Again this result has been anticipated by requiring the infinite differentiability of the test functions and using the "weak" convergence in Definition 2.6.

COROLLARY. A convergent series in $\mathfrak{J}'$ may always be differentiated term by term in the sense of generalized functions with the resulting series again converging in $\mathfrak{J}'$.

Proof. Consider the series $\sum_{n=1}^{\infty} \phi_n = \phi$ convergent in $\mathfrak{J}'$. By definition this means

$$\langle \phi , f \rangle = \lim_{n \to \infty} \langle S_N , f \rangle \quad \text{for all} \quad f \in \mathfrak{J}$$

where $S_N = \sum_{n=1}^{N} \phi_n$. Lemma 2.10 requires that $D^r S_N \xrightarrow[N \to \infty]{} D^r \phi$ for any multi-index r. Thus the series $\sum_{n=1}^{\infty} D^r \phi_n$ converges in the sense of Definition 2.6 to $D^r \in \mathfrak{J}'$.

EXAMPLE (e) A useful and important illustration of convergence for generalized functions is afforded by the sequence of Gaussian functions in one dimension

$$\phi_n = (n)^{\frac{1}{2}} \exp(-nx^2/4) \, / \, (4\pi)^{\frac{1}{2}} \qquad n = 1,2,3,\ldots \tag{2.21}$$

The ϕ_n have a maximum at $x = 0$ and fall to $1/e$ of this value for $x = \pm 2/n^{\frac{1}{2}}$. With increasing n, the ϕ_n become more sharply concentrated about the origin with increasing value. For any n however, the area under the curve is one.

For any test function $f \in \mathfrak{S}$, consider

$$\phi_n(f) - f(0) = (n/4\pi)^{\frac{1}{2}} \int_{-\infty}^{\infty} dx \, e^{-nx^2/4} \, [f(x) - f(0)] \tag{2.22}$$

and choose positive numbers $K > \delta > 0$ for which the range of integration in (2.22) is divided as follows

$$I_1 + I_2 + I_3 = (n/4\pi)^{\frac{1}{2}}\left[\int_0^\delta + \int_\delta^K + \int_K^\infty\right]dx\ e^{-nx^2/4}[f(x) - 2f(0) + f(x)] \ . \tag{2.23}$$

We now treat each term separately below.

(i) On $[K,\infty)$, change variables to $y = n^{\frac{1}{2}}x/2$, then

$$|I_3| \leq 1/\pi^{\frac{1}{2}} \int_{K n^{\frac{1}{2}}/2}^{\infty} dy\, e^{-y^2}\ |f(2y/-n^{\frac{1}{2}}) - 2f(0) + f(-2y/n^{\frac{1}{2}})| \ . \tag{2.24}$$

Since f is of rapid decrease, the expression involving f in (2.24) is bounded by a constant M uniformly in y and $n=1,2,3,\ldots$ Hence, we have the estimate

$$|I_3| \leq M/\pi^{\frac{1}{2}} \int_{n^{\frac{1}{2}}K/2}^{\infty} dy\ e^{-y^2}$$

which is a convergent integral decreasing uniformly with n. Thus, for any $\epsilon > 0$, we may choose $n \geq n_3(\epsilon)$ such that $|I_3| < \epsilon/3$.

(ii) On $[\delta,K)$, the mean-value theorem for integrals of continuous functions on a finite interval gives a bound

$$|I_2| \leq (n/4\pi)^{\frac{1}{2}}(K - \delta)\ |f(x_o) - 2f(0) + f(-x_o)|\ e^{-nx_o^2} \tag{2.25}$$

where $\delta \leq x_o \leq K$. Again since $f \in \mathcal{S}$, the term in (2.25) involving the test function is uniformly bounded on the finite interval, and for any $\epsilon > 0$, the exponential decrease of the right-hand side indicates that for $n \geq n_2(\epsilon)$, $|I_2| < \epsilon/3$.

(iii) Lastly, on $[0,\delta]$, we invoke the smoothness of the test functions in the form of a Taylor expansion about $x = 0$; namely,

$$f(x) - 2f(0) + f(-x) = x[f'(a_1 x) + f'(-a_2 x)] \quad 0 \le a_1, a_2 \le 1 \qquad (2.26)$$

where each $f'(ax)$ is again in $\mathcal{S}$, and hence, uniformly bounded on $[0,\delta]$ by M, say. Inserting (2.26) into the expression for I_1 and applying again the mean-value theorem, provides a bound

$$|I_1| \le (n/4\pi)^{\frac{1}{2}} M x_o \, e^{-nx_o^2/4}$$

with $0 \le x_o \le \delta$. If $x_o \ne 0$, the right-hand side decreases as n increases, so that if $\epsilon > 0$, we may choose $n \ge n_1(\epsilon)$ such that $|I_1| < \epsilon/3$.

Gathering together the three estimates just obtained, we have that for $n \ge \max[n_1, n_2, n_3]$ $|\phi_n(f) - f(0)| < \epsilon$ with $\epsilon > 0$ arbitrary. It then follows that

$$\lim_{n\to\infty} \phi_n(f) = f(0) = \delta_x(f) \qquad (2.27)$$

and $\phi_n \to \delta_x$ in $\mathcal{S}'$.

The same argument applies with minor modifications for less restrictive test functions f. The reader will have no difficulty in convincing himself that it is enough to have f obeying a weak smoothness condition near the origin, say a Lipschitz condition $|f(x) - f(0)| \le M|x|^{\alpha}$ with $0 < \alpha < 1$ and for $|e^{-\epsilon x^2} f(x)|$ to be uniformly bounded for any $0 < \epsilon < 1$. We shall not need such refinements, however.

From Lemma 2.10 we find in the sense of $\mathcal{S}'$ or $\mathcal{D}'$

$$D^r \phi_n \to D^r \delta_x . \qquad (2.28)$$

Particular cases for $r = 1,2$ are

$$-(n^3/16\pi)^{\frac{1}{2}} x\, e^{-nx^2/4} \to \delta_x' \;,\; (n^3/16\pi)^{\frac{1}{2}}(x^2/2 - 1)e^{-nx^2/4} \to \delta_x'' \;. \tag{2.29}$$

EXAMPLE (f) The previous example has a multi-dimensional analogue in R^n

$$\delta_x = \prod_{i=1}^{n} \delta_{x_i} = \lim_{N\to\infty} (N/4\pi)^{n/2} \exp(-N\|x\|^2/4) \tag{2.30}$$

where $\|x\|^2 = \Sigma_{i=1}^{n} x_i^2$. The convergence in $\mathcal{S}'$ or $\mathcal{D}'$ is established by using (2.27) in each variable separately.

PROBLEMS (5) Show that the convergence $\phi_n \to \delta_x$ in example (e) does not hold in $C^{\infty\prime}$.

(6) Show that the sequence $\{\phi_n\}$, defined by ϕ_n = zero for $x < 0$, x for $0 \le x \le 1/n$ and 1 for $x > 1/n$, converges in $\mathcal{S}'$ to $H(x)$. What is the limit of ϕ_n' ? Discuss the differences between pointwise convergence and convergence in the sense of generalized functions for this example.

(7) Show that $(\sin nx)/x \to \delta_x$ in $\mathcal{S}'$ or $\mathcal{D}'$. (<u>Hint</u>: Write $\int_{-\infty}^{\infty} dx\, (\sin nx/x)[f(x) - f(0)]$ as an integral over three regions as in (2.23). On the infinite region use the decrease of f and the oscillation of $\sin nx/x$ to obtain a decrease in n . For $[\delta,K]$, integrate by parts.)

2.5 DUAL SPACES

To gain some insight into the structure of distributions and their properties, it is necessary to carry further the discussion of section 1.2 on linear functionals. The results, to be obtained below, form useful knowledge concerning linear operators on linear spaces and will find specific use in this and the next section.

As in Chapter one, let V and V' denote two normed linear spaces and consider the collection of linear maps, $L(V,V')$, defined in section 1.2.

DEFINITION 2.11. A linear map $T \in (V,V')$ is said to be bounded if there exists a constant M such that $\|T(u)\|' \leq M\|u\|$ for all $u \in V$. The smallest number M, for which this is true, is called the norm of T and is written $\|T\|$.

Care should be taken to note that $\|T(u)\|'$ is the V' norm. The linearity of T allows us to write $\|T\|$ more compactly as

$$\|T\| = \sup_{\|u\|\leq 1} \|T(u)\|' \quad . \tag{2.31}$$

A notion, which is equivalent to boundedness for everywhere defined linear operators, is that of continuity.

DEFINITION 2.12. A linear map $T \in L(V,V')$ is said to be continuous if $\lim_{n\to\infty} \|T(u_n)\|' = 0$ for every sequence $\{u_n\} \subset V$ which tends to zero.

This definition is actually only applicable in a neighborhood of the origin in V. For any other point u_o, we make use of the linearity of T. Let $\{u_n\}$ be a sequence tending to u_o. Then $\{u_n - u_o\}$ tends to zero, and if T is continuous,

$\lim_{n\to\infty} \|T(u_n-u_o)\|' = \lim_{n\to\infty} \|T(u_n) - T(u_o)\| = 0$. Consequently, $\lim_{n\to\infty} \|T(u_n)\|' = \|T(u_o)\|'$.

PROPOSITION 2.13. A linear map $T \in L(V,V')$ is bounded if, and only if, it is continuous.

Proof. Suppose T is bounded and $\{u_n\}$ is a sequence tending to zero. Then $\lim_{n\to\infty} \|T(u_n)\|' \leq \lim_{n\to\infty} \|T\| \, \|u_n\| = 0$ and T is also continuous.

Conversely, assume T is continuous and that $\{u_n\}$ tends to zero. Then there exists an $\varepsilon > 0$ such that $\|u_n\| \leq \varepsilon$ implies $\|T(u_n)\|' \leq 1$. For any vector $u \in V$, we have $\|u\varepsilon/\|u\|\| \leq \varepsilon$ and hence $\|T(\varepsilon u/\|u\|)\|' \leq 1$. T is linear; therefore, $\|T(u)\|' \leq (1/\varepsilon)\|u\|$ and as u was arbitrary, $\|T\| \leq 1/\varepsilon$.

The construction used in the last part of the proof of this proposition will occur with some regularity in further proofs. Therefore, we shall refer to it as the "scaling trick".

With respect to the norm (2.31), the continuous linear functionals on V are of particular interest for us. Suppose $\{\phi_n\}$ is a sequence of such functionals. Then, by Proposition 2.13, each is also bounded. If we wish to express the idea that this sequence tends to zero, two alternatives present themselves. Either, in accordance with Definition 2.6, we may say

$$\phi_n \xrightarrow{w} 0 \quad \text{if} \quad \lim_{n\to\infty} \phi_n(u) = 0 \quad \text{for all} \quad u \in V \tag{2.32}$$

or

$$\phi_n \xrightarrow{s} 0 \quad \text{if} \quad \lim_{n\to\infty} \| \phi_n \| = 0 \ . \tag{2.33}$$

These two forms of convergence are very different in their implications and therefore should not be confused. The convergence in (2.32) is referred to as weak *-convergence, or weak convergence in the dual, and is the direct analogue for linear spaces of pointwise convergence for functions in the usual context. Relation (2.33) expresses the idea of uniform convergence and is called strong convergence in the dual.

DEFINITION 2.14. The set of continuous linear functionals on a normed space V with the strong convergence (2.33) will be referred to as the strong dual space of V, written V_s'.

Replacing strong convergence by weak convergence in this definition, we obtain the weak dual space of V, denoted V'. The spaces of generalized functions are the weak duals to $\mathfrak{J}$. Of course, $\mathfrak{J}$ is not a normed space and neither is $\mathfrak{J}'$. For weak convergence, the existence of a norm is not required. An important fact about V_s' is that it is a Banach space. This is not hard to prove and is left as an exercise.

PROBLEM (1) Let V be a Banach space. Show that V_s' is complete with respect to the norm (2.31).

The construction of test function spaces necessitated the building of additional structure beyond that of a normed space in the form of countably normed spaces and their inductive limits. In connection with the former, we found that

$$V \subset \dots \subset \hat{V}_{n+1} \subset \hat{V}_n \subset \dots \subset \hat{V}_1$$

(see (1.15)). Suppose $\{\phi_k\}$ is a sequence in $\hat{V}'_{n,s}$ which tends to zero strongly. Then for any $u \in V$, $\lim_{k\to\infty} |\phi_k(u)| \le \lim_{k\to\infty} \|\phi_k\| \|u\|_n = 0$ and $\{\phi_k\}$ also tends weakly to zero. From this follow the reverse inclusions

$$\hat{V}'_{1,s} \subset \dots \subset \hat{V}'_{n,s} \subset \hat{V}'_{n+1,s} \subset \dots \subset V' . \tag{2.34}$$

More is, in fact, true.

PROPOSITION 2.15. Let V be a countably normed space. Then $V' = \bigcup_{n=1}^{\infty} \hat{V}'_{n,s}$.

Proof. In the course of deriving (2.34), we verified that $\bigcup_{n=1}^{\infty} V'_{n,s} \subset V'$, so only the reverse inclusion is needed. This is a simple use of the scaling trick. For $\phi \in V'$, as the neighborhoods (1.16) form a base for the topology of V , there exists $\varepsilon > 0$ and a positive integer n for which $\|u\|_n < \varepsilon$ implies $|\phi(u)| \le 1$. Hence, $\|\phi\|_n \le 1/\varepsilon$ and $\phi \in \hat{V}'_{n,s}$.

EXAMPLES (a) By Theorem 1.24, $\mathcal{D}(K)$ is a complete countably normed space for which Proposition 2.15 gives $\mathcal{D}'(K) = \bigcup_{m=0}^{\infty} \mathcal{D}^{(m)}(K)'_s$.

(b) The inclusions $\mathcal{D} \subset \mathcal{S} \subset C^\infty$ imply $C^{\infty\prime} \subset \mathcal{S}' \subset \mathcal{D}'$. Clearly, $e^x \in \mathcal{D}'$ but not $\mathcal{S}'$, while $x^m \in \mathcal{S}'$ but not in $C^{\infty\prime}$. All of the generalized functions among $\mathfrak{J}'$ are special classes of distributions.

PROBLEM (2) Consider the Banach spaces $C_0^0(R)$ and $\mathcal{D}^0(K)$ where K is a finite interval. Define a linear functional on $\mathcal{D}^0(K)$ which is not in $C_0^0(R)'$. (Hint: Consider example 1.2 (b).)

The relation between the weak dual V' of a countably normed space and the strong dual spaces $\{\hat{V}'_{n,s}\}$ given in proposition 2.15 is the key to much of the more technical portions of distribution theory. We shall discuss two such features: the completeness of $\mathcal{D}'$ and, in section 2.6, support properties of distributions. For the reader who is less inclined toward technicalities, these may be omitted at a first reading and he might pass directly to section 2.7 noting only the statements of Theorems 2.17, 2.18 and 2.19.

The proof of completeness for the weak dual of a complete countably normed space will require a lemma: the Baire category theory for complete metric spaces. This is one of the famous and important results in functional analysis (see problem 3).

LEMMA 2.16. Let V be a complete, countably normed space with weak dual V'. Suppose $\{\phi_k\} \subset V'$ converges weakly; that is, $\lim_{k\to\infty} \phi_k(u) = \phi(u)$ exists for each $u \in V$. Then the set $M = \{u \in V \mid |\phi_k(u)| \leq 1 \text{ for all } k=1,2,\ldots\}$ contains a neighborhood of the form (1.16).

Proof. Consider a sequence $\{u_p\} \subset M$ which converges to u in V. Since each ϕ_k is continuous, $|\phi_k(u)| = \lim_{p\to\infty} |\phi_k(u_p)| \leq 1$ and M is closed.

Next, pick any $u \in V$ and note that

$|\phi_k(u)| \le |\langle \phi - \phi_k, u\rangle| + |\phi(u)|$. Choose $k \ge k_o(\epsilon, u)$ so that $\sup_{k \ge k_o} |\langle \phi - \phi_k, u\rangle| < \epsilon$. Then

$$\sup_{k=1,2,\ldots} |\phi_k(u)| \le \sup_{k<k_o} [\epsilon + |\phi_k(u)| + |\phi(u)|] = m(u)$$

with the result that $u/m(u) \in M$. This means that upon multiplying all vectors in M by a suitably large constant, we may incorporate an arbitrary vector u in a set of the form $M_n = \{n\,u|\, u \in M\}$, n a positive integer. Consequently, $V = \bigcup_{n=1}^{\infty} M_n$.

The lemma will now be proved by obtaining a contradiction on the assumed completeness of V . Suppose each M_n contains no non-empty open set of V . Then for each n , the complement, M_n^c , is an open set which is dense in V . To see this, suppose otherwise. Namely, there exists $u \in (M_n^c)^c$ = interior of M_n which would be a non-empty open subset of M_n contradictory to our hypothesis for M_n . Starting with M_1 , choose a neighborhood of the form $N_{\frac{1}{2},1}(u_1)$ contained in M_1^c . This is a "sphere" of radius $1/2$ for the 1-norm with center at u_1 . Generally, for each integer n we select a neighborhood $N_{1/2^n,\,n}(u_n)$ contained in M_n^c and in the previous neighborhood $N_{1/2^{n-1},n-1}(u_{n-1})$. This gives a sequence of "nested spheres", each one lying outside M_n . Consider the sequence $\{u_n\}$ formed from their centers. This is a Cauchy sequence by virtue of the estimate

$$\|u_p - u_q\|_n \le \|u_p - u_q\|_q \le 1/2^q \quad \text{for} \quad p \ge q \ge n$$

and hence has a limit $u_o \in V$. Moreover, for any $\epsilon > 0$

$$\|u_o - u_n\|_n \le \|u_o - u_p\|_n + \|u_p - u_n\|_n < \varepsilon + 1/2^n$$

with p sufficiently large. We now have arrived at our contradiction; namely, $u_o \in N_{\cdot,n}$ for each n, requires $u_o \in M_n^c$ also for each n. Then $u_o \notin V$!

From the contradiction in the previous paragraph, some M_n must contain a non-trivial open set of V. Choose a suitable vector v such that $v + M_n$ contains the origin and hence an open set about this point. Now appeal to the fact that (1.16) form a base so that $v + M_n$ contains a neighborhood of this form containing zero. Finally, by scaling with a suitable constant a, this neighborhood can be "shrunk" into M since $a(v + M_n) \subset M$. This is our result.

PROBLEMS (3) A set X is called nowhere dense if its closure contains no non-empty open set. Let X be a complete metric space and $\{X_n\}$ a family of nowhere dense subsets. Show that the proof of Lemma 2.16 demonstrates that $X = \bigcup_{n=1}^{\infty} X_n$ is impossible. This is the Baire category theorem.

(4) Let V be a complete, countably normed space. Show that V is a complete metric space with respect to $d(u,v) = \Sigma_{n=1}^{\infty} (1/2^n)\|u - v\|_n / (1 + \|u - v\|_n)$.

Lemma 2.16 will now be used in the proof of the following result.

THEOREM 2.17. The spaces of generalized functions $\mathcal{J}'$ are complete.

Proof. The proof will actually be given only for $\mathcal{D}'$ and $\mathcal{S}'$, in which case, it may be reduced to a common result by the following observation. The form of convergence in $\mathcal{D}$ (see below Theorem 1.26) requires that $\phi \in \mathcal{D}'$ if, and only if, $\phi \in \mathcal{D}'(K)$ for each compact set $K \subset R^n$. Our task is then to show completeness for $\mathcal{D}'(K)$ and $\mathcal{S}'$, both of which are complete, countably normed spaces.

Consider a Cauchy sequence $\{\phi_p\} \subset V'$, the weak dual of a complete, countably normed space V. Then given $\varepsilon > 0$, $n_o(\varepsilon,u)$ can be chosen such that $|\phi_p(u) - \phi_q(u)| < \varepsilon$ for $p \geq q \geq n_o$. The sequence $\{\phi_p(u)\}$ of complex numbers has a limit, say $\phi(u)$. This defines a linear functional since

$$\lim_{p\to\infty} \phi_p(\alpha u + \beta v) = \alpha \lim_{p\to\infty} \phi_p(u) + \beta \lim_{p\to\infty} \phi_p(v)$$

or $\phi(\alpha u + \beta v) = \alpha\phi(u) + \beta\phi(v)$. For ϕ to be in V', it must be shown that it is weakly continuous. In this regard, consider $\lim_{k\to\infty} \phi(u_k) = \lim_{k\to\infty} \lim_{p\to\infty} \phi_p(u_k)$ where $u_k \to 0$. If this convergence were known to be uniform, the order of the limit could be interchanged and we could conclude $\lim_{k\to\infty} \phi(u_k) = \lim_{p\to\infty} \lim_{k\to\infty} \phi_p(u_k) = 0$ from the continuity of ϕ_p. It is precisely Lemma 2.16 which leads to the necessary uniformity.

Let $M = \{u \in V \mid |\phi_p(u)| \leq 1 \quad p = 1,2,\ldots\}$ be the set introduced in Lemma 2.16. As shown there, M contains a neighborhood of the form (1.16); that is, there exists $\delta > 0$ and an integer n for which $\|u\|_n < \delta$ implies $|\phi(u)| = \lim_{p\to\infty} |\phi_p(u)| \leq 1$. The scaling trick once again leads to $\phi \in \hat{V}'_{n,s}$ and $\phi \in V'$ by Theorem 2.15.

2.6 DISTRIBUTIONS OF COMPACT SUPPORT

In advanced works on the subject of generalized functions, one will find many results pertaining to specific representations for these linear functionals in terms of integrals, or more precisely, measures and derivatives of continuous functions. Needless to say, almost all of this work lies outside the scope of this book as it requires a substantial amount of functional analysis. This body of knowledge is at the heart of the theory of distributions as applied to modern studies in the theory of partial differential equations. Our excursion into such results will be brief and will deal only with the most accessible ideas.

Let ϕ be a distribution with compact support $K \subset R^n$ and for some $x_o \in K$ define

$$r_o = r(x_o) = \sup_{x \in K} \|x - x_o\| \ . \tag{2.35}$$

Since K is compact, the set of real numbers $\{\|x - x_o\| \mid x \in K\}$ is bounded and the supremum in (2.35) exists. Geometrically, $r(x_o)$ is the maximum distance of x_o from the boundary (∂K) of K. We shall also need to use the function $\theta_1(\|x\|/(r(x_o) + \delta))$ introduced in (1.25). Here $\delta > 0$ and this function lies in $\mathcal{D}(B_o + \delta)$, where the sphere with center at x_o and radius $r(x_o) + \delta$ is denoted by $B_o + \delta$.

For any $f \in \mathcal{D}$, the linearity of ϕ gives

$$\phi(f) = \langle \phi, f\theta_1(\|x\|/r_o + \delta)\rangle + \langle \phi, [1 - \theta_1(\|x\|/(r_o + \delta))]f\rangle$$

$$= \langle \phi, \theta_1(\|x\|/(r_o + \delta))f\rangle \tag{2.36}$$

since $\operatorname{supp} \phi \cap \{x \in R^n \mid \|x - x_o\| > r_o + \delta\}$ is empty. From this relation ϕ can be defined uniquely as an element of $\mathcal{D}'(B_o + \delta)$, and hence, it follows from example 2.5 (a) that some positive integer m exists for which $\phi \in \mathcal{D}^{(m)}(B_o + \delta)_s'$. This last fact leads to our first "structure" theorem for distributions.

THEOREM 2.18. Let $\phi \in \mathcal{D}'$ with compact support K. Then for any $\delta > 0$, there exists a positive integer m and constant C such that for all f

$$|\phi(f)| \leq C\| f \|_{B_o + \delta, m}$$

$$\leq C \sup_{x \in B_o + \delta, \ |r| \leq m} |D^r f(x)| .$$

The integer m depends only upon K and δ, not on the test function f.

The proof of this result needs only the form of the norm in (1.23) added to our remarks above.

A special case of Theorem 2.18 is illustrated when ϕ has the single point x_o as support. The structure of ϕ may be completely exposed in this case.

THEOREM 2.19. For $\phi \in \mathcal{D}'$ with $\operatorname{supp} \phi = \{x_o\}$, there exists a positive integer m and constants c_r, $|r| = 0,1,\ldots,m$ such that

$$\phi = \sum_{|r| \leq m} c_r D^r \delta_{x-x_o} , \text{ uniquely.}$$

Proof. ϕ clearly has compact support and Theorem 2.18 can be applied to indicate that $\phi \in \mathcal{D}^{(m)'}(B_\delta)_s$ where B_δ is a sphere of radius δ around x_o. Moreover, (2.36) yields $\phi(f) = \langle \phi , \theta_1(\| x \| / \delta)f \rangle$ for any $f \in C^\infty$. Denote $g(x) = \theta_1(\|x\|/\delta)f(x) \in \mathcal{D}(B_\delta)$, then expand in a Taylor's series of order m about x_o; that is,

$$g(x) = \sum_{|r| \leq m} D^r g(x)/r! \Big|_{x=x_o} (x-x_o)^r + g_R(x) \tag{2.37}$$

where the remainder g_R has the property that

$$g_R \in \mathcal{D}(B_\delta) \quad D^r g_R(x) \Big|_{x=x_o} = 0 \quad |r| \leq m . \tag{2.38}$$

Substituting (2.37) into (2.36), we find

$$\phi(f) = \phi(g) = \sum_{|r| \leq m} D^r g \Big|_{x=x_o} \langle \phi , (x - x_o)^r \rangle / r! + \phi(g_R) .$$

Yet, $D^r g \Big|_{x=x_o} = (-1)^{|r|} \langle D^r \delta(x - x_o) , g \rangle$, and we may rewrite the last expression in the form

$$\phi(f) = \sum_{|r| \leq m} c_r \langle D^r \delta(x - x_o) , g \rangle + \phi(g_R)$$

where $c_r = (-1)^{|r|} \langle \phi , (x - x_o)^r \rangle / r!$. Finally, applying Theorem 2.18, we can obtain a bound on the remainder term. In fact, by choice of the integer m, there exists a constant C for which $|\phi(g_R)| \leq C \sup_{\|x-x_o\|<\delta,\ |r| \leq m} |D^r g_R(x)|$. The conditions of (2.38) show that for δ small enough, $|\phi(g_R)|$ can be made arbitrarily small.

EXAMPLE. In order to see the reason behind Theorem 2.19, consider the linear functional $\phi = \Sigma_{r=0}^{\infty} (-1)^r \delta^{(r)}(x)$ on elements from C^{∞}, in particular the test function e^x. We find $\phi(e^x) = \Sigma_{r=0}^{\infty} (1)^r$, a series which certainly does not converge.

The last example raises an interesting question; namely, just what sort of distributions belong to $C^{\infty\prime}$? The relation in (2.36) implies that this set contains the ones with compact support; in fact, this is all of $C^{\infty\prime}$.

THEOREM 2.20. For $\phi \in C^{\infty\prime}$, supp ϕ is compact.

Proof. Given any $f \in C^{\infty}$, consider $f = \theta(x/n)f + [1 - \theta(x/n)]f$ where θ is the function constructed in problem 1.7 (2) and n is a positive integer. It is easy to check that the sequence $f_n = [1 - \theta(x/n)]f$ tends to zero in C^{∞}. Thus, we have $\phi(f_n) \to 0$ as $n \to \infty$.

Let us suppose, however, that ϕ does not have compact support. Then $\phi(f_n) \neq 0$ for each integer n and, in fact, we may choose f so that $\inf_n \{\phi(f_n)\} > 0$. This contradicts the continuity of ϕ above.

As an addendum to this section, one might ask whether a bound of the form appearing in Theorem 2.18 holds more generally when the distribution does not have compact support. The answer is affirmative, and the conditions for its validity may be seen by retracing the argument leading to this result. Suppose we examine $\phi(f)$ for any $f \in \mathcal{D}$. If we choose $K = \text{supp } f$, the desired estimate results by the same reasoning as before. The essential difference is that the index m depended upon K and δ ; it must now depend on the test function f .

THEOREM 2.21. For $\phi \in \mathcal{D}'$ and $f \in \mathcal{D}$ there exists a constant C and positive integer m such that

$$|\phi(f)| \leq C \sup_{|r| \leq m,\ x_0 \in B_0 + \delta} |D^r f(x)| . \tag{2.39}$$

If m may be chosen independently of f , we say that ϕ is of finite order and define the order of ϕ to be the smallest positive integer m for which (2.39) holds. Theorem 2.18 may then be restated to say that every distribution of compact support has finite order.

2.7 DIRECT PRODUCTS AND CONVOLUTIONS

The next set of rules to be built from those for ordinary functions concerns products and convolutions of generalized functions. Given two generalized functions, ϕ and ψ , we ask under what conditions is their product $\phi\psi$ again a generalized function. If ϕ and ψ are in the same variables, the answer is almost always no. Only for special generalized functions can the product $\phi(x)\psi(x)$ have a meaning within the same context as for the individual constituents. The reason is simply that for any test function f , $\langle \phi\psi , f \rangle = \int dx\ \phi(x)\psi(x)f(x)$, is well-defined only if $\psi(x)f(x)$ is a test function for ϕ . In this sense, generalized functions do not always lie in the multiplier space for ϕ. If, however, ϕ and ψ are in different variables, then $\phi(x)\psi(y)$ will be a generalized function in two variables. When this is the case, we call $\phi(x)\psi(y)$ the direct product of the generalized functions ϕ and ψ, and denote it by $\phi \otimes \psi$. To summarize these remarks, we make the following definition.

DEFINITION 2.22. For $\phi \in \mathcal{J}'(R^k)$ and $\psi \in \mathcal{J}'(R^\ell)$, their direct product $\phi \otimes \psi$ defines a generalized function in $\mathcal{J}'(R^n)$, $n = k + \ell$, by means of the relation

$$\phi \otimes \psi(f) = \langle \phi , \psi(f) \rangle = \langle \psi , \phi(f) \rangle \tag{2.40}$$

for all $f \in \mathcal{J}(R^n)$.

The notation in (2.40) requires some explanation. We have used $\psi(f)$ to indicate the result of applying ψ to the test function f. If we denote a vector in R^n by $x = (y_1, \ldots, y_k, z_1, \ldots, z_\ell)$, it follows that $\psi(f) = \int dz\ \psi(z)\ f(y, z)$. Similarly, $\phi(f) = \int dy\ \phi(y)\ f(y, z)$. These last integrals are to be understood only in a symbolic sense.

A natural consequence of this definition is the commutativity of the direct product

$$\phi \otimes \psi = \psi \otimes \phi \ . \tag{2.41}$$

It is an easy matter to extend the definition to many factors through the formula

$$\overset{p}{\underset{j=1}{\otimes}} \phi_j(f) = \langle \phi_1 , \overset{p}{\underset{j=2}{\otimes}} \phi_j(f) \rangle$$

$$= \langle \phi_1 , \langle \phi_2 , \langle \ldots , \langle \phi_p , f \rangle \rangle \ldots \rangle \rangle \rangle \tag{2.42}$$

where $f \in \mathfrak{J}(R^n)$, $n = n_1 + n_2 + \ldots + n_p$, and each $\phi_j \in \mathfrak{J}'(R^{n_j})$. The direct product so defined is commutative. Consequently, if $\pi = (\pi_1, \pi_2, \ldots, \pi_p)$ is any permutation of $(1,2,\ldots,p)$, then

$$\overset{p}{\underset{j=1}{\otimes}} \phi_j = \overset{p}{\underset{j=1}{\otimes}} \phi_{\pi_j} \quad . \tag{2.43}$$

PROBLEMS (1) Consider the test function $f(x) = \prod_{j=1}^{n} g(x_j)$, $g \in \mathfrak{J}(R)$. Show that if $\phi_j \in \mathfrak{J}'(R)$, then $\overset{n}{\underset{j=1}{\otimes}} \phi_j(f) = \prod_{j=1}^{n} \phi_j(g)$.

(2) Verify the relation $\delta_{x_1} \otimes \delta_{x_2} = \delta_{(x_1,x_2)}$ using induction to deduce $\overset{n}{\underset{j=1}{\otimes}} \delta_{x_j} = \delta_x$.

One of the most important and useful applications of the direct product of two generalized functions is that it enables us to define the idea of convolution for such quantities. Recall that for two functions, f and g , let us say in $\mathfrak{S}$, their convolution is the expression $f*g$ given by

$$f*g(x) = \int dy\ f(x-y)\ g(y) = \int dy\ f(y)\ g(x-y)$$
$$= g*f(x) \quad .$$

One may easily check that this is well-defined and is again in $\mathfrak{S}$. One consequence is that we expect the relation

$$\int dx\ h(x)\ f*g(x) = \int dx\, dy\, h(x+y)\, f(x)\, g(y) = \langle f \otimes g(x,y), h(x+y) \rangle$$

will exist for any reasonable function h . For generalized functions, it is desirable that similar properties should be true.

DEFINITION 2.23. Consider $f \in \mathfrak{J}$, and let ϕ and ψ be two generalized functions. Their convolution $\phi * \psi$ is a generalized function in $\mathfrak{J}'$ defined by $\phi * \psi(f) = \langle \phi \otimes \psi , f(x+y) \rangle$, whenever this expression exists.

It should be pointed out immediately that this definition will not hold for arbitrary generalized functions ϕ and ψ . It will surely work when $f(x+y)$ is a test function in the variables (x, y) appropriate to ϕ and ψ . Smoothness, clearly, presents no problem as $f \in C^{\infty}$ implies that $f(x+y)$ is C^{∞} in the variables x and y separately. The main difficulty with Definition 2.23 of a convolution concerns the support properties of $f(x+y)$ in the two sets of variables. For example, suppose $f \in \mathfrak{D}$, then $f(x+y)$ need not have compact support in x and y even though this is the case for a fixed value of either one. This indicates that $\phi * \psi$ will not, in general, be a distribution without further specialization. The two lemmas below provide sufficient conditions under which $\phi * \psi$ is defined, yet do not exhaust the possibilities.

LEMMA 2.24. Let $\phi \in C^{\infty\prime}$ and $\psi \in \mathfrak{J}'$. Then $\phi * \psi \in \mathfrak{J}'$.

Proof. Using the commutativity of the direct product, definition 2.23 yields

$$\langle \phi * \psi , f(x+y) \rangle = \langle \psi , \langle \phi_y , f(x+y) \rangle \rangle$$

and our result follows once it is established that $f \in \mathfrak{J}$ implies $\langle \phi_y , f(x+y)\rangle(x) \in \mathfrak{J}$.

(i) As already remarked above, this expression is C^∞ in x for any of the test functions from $\mathfrak{J}$ since

$$D^r \langle \phi_y , f(x+y)\rangle = \langle \phi_y , D^r f(x+y)\rangle \quad . \tag{2.44}$$

This takes care of the case $\psi \in C^{\infty\prime}$.

The remaining two cases depend upon the result of Theorem 2.20 that ϕ has compact support.

(ii) When $f \in \mathcal{D}$, $\langle \phi_y , f(x+y)\rangle$ vanishes unless $y \in \operatorname{supp} \phi$ and $x+y \in \operatorname{supp} f$. This determines a compact set of values for the support of $\phi_y(f(x+y))$ in x . This argument plainly extends without change to any derivative of the form (2.44). Thus, the convolution $\phi * \psi$ exists for $\psi \in \mathcal{D}'$.

(iii) The last case to be considered is when $f \in \mathcal{S}$. Once again, the compact support of ϕ implies that $\langle \phi_y , f(x+y)\rangle$ vanishes unless $y \in \operatorname{supp} \phi$. Now, $f \in \mathcal{S}$ implies that $f(x+y)$ is likewise of rapid decrease in x for bounded y . More preciscly, for any multi-index r and integer N , the expression $(1+\|x\|^2)^N D_x^r f(x+y) \in C^\infty$ in the variable y for all bounded values of y . The result being that

$$(1+\|x\|^2)^N D^r \langle \phi_y , f(x+y)\rangle = \langle \phi_y , (1+\|x\|^2)^N D^r f(x+y)\rangle$$

is finite. From this we conclude that when $f \in \mathcal{S}$, $\langle \phi_y , f(x+y)\rangle$ is also in $\mathcal{S}$ in the variable x . This finishes the proof of Lemma 2.24.

PROBLEM (3) Show that $\phi * \delta = \phi$ for all $\phi \in \mathcal{J}'$, and thus, deduce the relations $\phi * \delta * \psi = \phi * \psi * \delta = \phi * \psi$ if $\psi \in C^{\infty\prime}$.

The second condition for which we show the existence of the convolution will be stated only in the case of one dimension.

LEMMA 2.25. Let $\operatorname{supp} \phi \subset (a, \infty)$ and $\operatorname{supp} \psi \subset (b, \infty)$ where a and b are real numbers. If ϕ and $\psi \in \mathcal{D}'$, then $\phi * \psi \in \mathcal{D}'$.

Proof. Again we consider the test function given by $\langle \phi_y, f(x+y) \rangle$ in the variable x. For $f \in \mathcal{D}$ there will be a non-vanishing contribution when $y \in (a, \infty)$; namely, for x contained in a set of the form $(-\infty, c)$ for some real number c. Such is true also for any derivative (2.44). The restriction on the assumed form for the support of ψ indicates that only those values of x contribute from an interval of the form (b, c) (or (c, b) if $c < b$). This is a finite interval. Therefore $\phi * \psi(f)$ certainly exists.

The essential feature in Lemma 2.25 was that both distributions have supports which are bounded below. The form of the convolution does the rest.

PROBLEMS (4) Rework the proof of Lemma 2.25 for the case when $\operatorname{supp} \phi$ and $\operatorname{supp} \psi$ are bounded from above. Is the lemma still true when $\operatorname{supp} \phi$ is bounded from below, but $\operatorname{supp} \psi$ is bounded from above?

(5) Generalize Lemma 2.25 to the case of two variables. State the result in n-dimensions.

Some useful by-products of these last two lemmas are given as corollaries below. In each case the proofs follow the same reasoning as above.

COROLLARY 1. For $\phi \in \mathfrak{J}'$ and $f \in \mathfrak{J}$, $\phi * f$ exists and is C^∞.

COROLLARY 2. For $\phi \in C^{\infty\prime}$ and $f \in \mathfrak{J}$, $\phi * f \in \mathfrak{J}$.

Given the existence of the convolution $\phi * \psi$ of two generalized functions, we may easily verify

$$D^r(\phi * \psi) = (D^r \phi) * \psi = \phi * (D^r \psi) \quad . \tag{2.45}$$

Indeed, for an appropriate test function f

$$\begin{aligned} D^r(\phi * \psi)(f) &= (-1)^{|r|} \phi * \psi(D^r f) = (-1)^{|r|} \langle \phi , \psi(D^r f) \rangle \\ &= \langle \phi , (D^r \psi)(f) \rangle = \langle \phi * (D^r \psi) , f \rangle \\ &= \langle \psi , (D^r \phi)(f) \rangle = \langle (D^r \phi) * \psi , f \rangle \quad . \end{aligned}$$

The reader is advised to trace carefully each of these steps, then justify them by using the commutativity of the convolution and the appropriate smoothness in the variables for which the differentiations are carried out. Similar relations to (2.45) hold for multiple convolutions whenever the latter exist.

The following problem frequently arises in applications. Given a sequence $\{\phi_n\}$ of generalized functions converging to $\phi \in \mathfrak{J}'$, form the convolutions $\phi_n * \psi$ which are assumed to exist for each n. Does the limit $\lim_{n \to \infty} \phi_n * \psi$ exist? The answer is to be found in the next formal proposition.

PROPOSITION 2.26. Let $\phi_n \to \phi$ in $\mathfrak{J}_1'$ and suppose $\psi * \phi_n$ exists in $\mathfrak{J}_2'$ for each integer n as in Lemmas 2.24 and 2.25. Then $\psi * \phi_n \to \psi * \phi \in \mathfrak{J}_2'$.

Proof. Consider $f \in \mathfrak{J}_2$. Then Definition 2.23 produces the relations

$$\psi * \phi_n(f) = \langle \psi , \langle \phi_{n,y} , f(x+y) \rangle\rangle \to \langle \psi , \langle \phi_y , f(x+y) \rangle\rangle$$

Since $\phi \in \mathfrak{J}_1'$, the hypothesis that the $\psi * \phi_n$ exist provides that $\langle \phi_y , f(x+y) \rangle$ satisfies the conditions of either Lemma 2.24 or 2.25 for a test function for ψ. Thus, $\psi * \phi_n(f) \to \psi * \phi(f)$.

In the next section we shall give several examples of this last result for the smoothing, or regularization, of generalized functions.

2.8 APPROXIMATION BY FUNCTIONS IN $\mathfrak{D}$

Although the idea of a generalized function has been made precise in the preceeding sections, we have not paid attention as to how such mught be realized. Example (e) of section 2.4 suggests that test functions could be used to approximate generalized functions. This is, in fact, the case, and suitable conditions are examined in this section.

Our first result in this direction shows that any continuous function in R^n tending to zero at infinity can be uniformly approximated by C^∞ functions of compact support. This hints that the appropriate test functions to be used are those in $\mathfrak{D}$.

THEOREM 2.27. The functions in $\mathcal{D}(R^n)$ are dense in $C_o(R^n)$.

<u>Proof</u>. We first observe that functions in $\mathcal{D}^o(R^n)$ are dense in $C_o(R^n)$. To see this, pick any $g \in \mathcal{D}^o$ which is one on the sphere $\|x\| \le 1$. Then, for any $f \in C_o$, $\lim_{n\to\infty} \sup_{x\in R^n} |f(x) - g(x/n)f(x)| = 0$. Thus, it is enough to show that any continuous function of compact support may be approximated by elements in $\mathcal{D}$ with respect to the supremum norm (1.19).

Consider the function $\theta_n(x) = c\ \theta_1(x\ n)$ where $n = 1,2,\ldots$ and $\int dx\ \theta_n(x) = 1$ (see problem 1.7 (2)). For any $f \in \mathcal{D}^o$, construct the sequence

$$f_n(x) = \int dy\ \theta_n(x-y)f(y) = (\theta_n * f)(x) \quad .$$

Clearly, for each multi-index r, $D^r f_n(x) = [(D^r\theta_n) * f](x)$ exists as a uniformly convergent integral. The integration is over a bounded region of those values of y for which $\{\|x-y\| \le 1/n\} \cap \operatorname{supp} f$ is not empty. Since $\operatorname{supp} f$ is compact, this is possibly only for a bounded region in X. Hence, each $f_n \in \mathcal{D}$.

The sequence $\{f_n\}$ converges to f uniformly as

$$\sup_x |f(x) - f_n(x)| = \sup_x \left|\int dy\ \theta_n(y)f(x) - \int dy\ \theta_n(x-y)f(y)\right|$$

$$\le \int dy\ \theta_n(y) \sup_x |f(x) - f(x-y)| \ .$$

Since $\|y\| \le 1/n$ in the integrand, by choosing n large enough, we can make $\sup_x |f(x) - f(x-y)|$ arbitrarily small and $\|f - f_n\|$ tends to zero as $n \to \infty$. This is the result we set out to prove.

COROLLARY 1. The functions $\mathcal{D}(R^n)$ are dense in $C^\infty(R^n)$.

<u>Proof</u>. It is enough to remark that convergence in C^∞, as given by Definition 1.18, requires uniform convergence of both functions and their derivatives on compact sets. However, this is exactly the form of convergence proved in Theorem 2.27 above.

COROLLARY 2. The functions $\mathcal{D}(R^n)$ are dense in $\mathcal{S}(R^n)$.

PROBLEM (1) Give a proof of Corollary 2.

The last two corollaries show that $\mathcal{D}$ is dense in the spaces of test functions which we consider at present. This is also true for the corresponding generalized functions.

THEOREM 2.28. Every generalized function $\phi \in \mathfrak{T}'$ is the limit of a sequence $\{f_n\}$ of functions from $\mathcal{D}$. The convergence is that appropriate to $\mathfrak{T}'$; that is $\langle f_n, f\rangle \to \langle \phi, f\rangle$ for all $f \in \mathfrak{T}$.

<u>Proof</u>. Let $\phi \in \mathfrak{T}'$ and suppose $\{g_n\}$ is a sequence of functions in $\mathcal{D}$ such that $g_n \to \delta$ in $C^{\infty\prime}$. Then by Corollary 1 to Lemma 2.25, each $\phi * g_n$ is C^∞ and $\phi * g_n \to \phi * \delta = \phi$. This last convergence takes place in $\mathfrak{T}'$ from Proposition 2.26. Finally to obtain the result stated in the theorem, choose a further sequence of functions $\{h_{n,m}\}$ in $\mathcal{D}$ which converge in the sense of $\mathfrak{T}'$ to $\phi * g_n$. The diagonal sequence $\{f_n = h_{n,n}\}$ is the one required.

The process of approximating ϕ by smooth functions $\{\phi * g_n\}$ or

$\{f_n\}$ is frequently called regularization, and each member of the approximating sequence a smoothing of ϕ. The practical implication of Theorem 2.28 is that every generalized function may be realized to be an arbitrary degree of approximation in the distribution topology by C^∞ functions of compact support. Needless to say, this is not always a convenient or simple construction to make.

PROBLEM (2) Find a sequence $\{g_n\}$ in $\mathfrak{D}$, such that $g_n \to \delta$ in each one of the spaces $\mathfrak{J}'$. (<u>Hint</u>: Consider the functions θ_n used in the proof of Theorem 2.27.)

3. FOURIER ANALYSIS

The basic operations for generalized functions are now applied to study the convergence of Fourier series to distributions in $\mathcal{S}'$. The Fourier transform is defined for test functions, then extended to generalized functions by duality. Some elementary characterizations of the spaces so obtained are given.

3.1 FOURIER SERIES

Before beginning, it might be as well to recall some of the classical results relating to the convergence of a Fourier series. We must presume some knowledge on the part of the reader concerning the elementary properties of convergence for a series of functions. In the case of Fourier series, this may be found in _An Introduction to Fourier Series and Integrals_, by R. Seeley, [7].

Consider a function $f(x)$ defined on the interval $0 \leq x \leq 2\pi$, and outside this region by periodicity, $f(x+2\pi) = f(x)$. A trigonometric series of the form

$$f(x) = a_o/2 + \sum_{n=1}^{\infty} a_n \cos nx + b_n \sin nx \tag{3.1}$$

is called a Fourier series for $f(x)$ on the interval $[0,2\pi]$. The coefficients a_n, b_n are given formally by

$$a_n = 1/\pi \int_0^{2\pi} dx\, f(x) \cos nx \quad b_n = 1/\pi \int_0^{2\pi} dx\, f(x) \sin nx \quad (3.2)$$

A first reasonable requirement for the coefficients in (3.1) to be given by (3.2) might be that the integrals should exist. In particular, this will be the case when f is absolutely integrable.

The above, in itself, is not a very stringent condition. In applications, however, it may be desirable to differentiate (3.1) term by term; for example, when (3.1) is used to solve a differential equation with periodic boundary conditions. An early result in the theory of Fourier series, [8], permits such term by term differentiation if $f(x)$ is continuous for all x in the closed interval $[0,2\pi]$ and $f'(x)$ has only a finite number of discontinuities. Both f and f' are required to be bounded on the given interval. A simple example indicates the weakness of this statement, and how it might be generalized to cover a wider class of functions.

Let $f(x) = x$ on the interval $0 < x < 2\pi$, and extended to the whole line as in Figure 3.1. The Fourier series for this function is

$$f(x) = \pi - 2 \sum_{n=1}^{\infty} (\sin nx)/n \qquad (3.3)$$

which is convergent on the open interval. This is not the case at the end-points, the series being a sum of terms which are zero and π while $\lim_{x\to 0^+} f = 0$, $\lim_{x\to 2\pi^-} f = 2\pi$. We should notice, nonetheless, that (3.3) does converge to the mean $\frac{1}{2}[f(0^+) + f(2\pi^-)]$. Such a situation is again covered

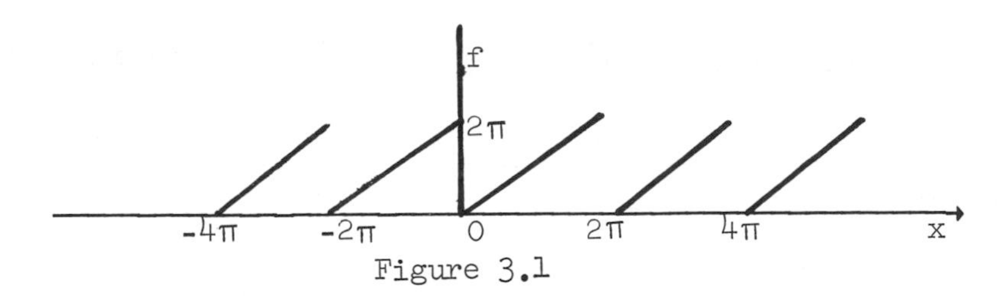

Figure 3.1

by a classical theorem.

THEOREM 3.1. Let f be defined on $(0,2\pi)$, periodic and piecewise differentiable. Then at each point x, the partial sums $S_N(x) = \Sigma_{n=-N}^{N} c_n e^{inx}$ converge to $\frac{1}{2}[f(x+) + f(x-)]$.

In stating this result, the complex form of (3.1) has been used in which

$$f(x) = \sum_{n=-\infty}^{\infty} c_n e^{inx} \qquad c_n = 1/2\pi \int_0^{2\pi} dx\, f(x)\, e^{-inx} \tag{3.4}$$

and $c_n = \frac{1}{2}(a_n - ib_n)$ for $n \geq 0$, $\frac{1}{2}(a_n + ib_n)$ for $n < 0$.

The series obtained by differentiating (3.3) is $-2\,\Sigma_{n=1}^{\infty} \cos nx$ which is not convergent in the usual sense. Yet, on the open interval $f' = 1$. A further notion of convergence for (3.1) is introduced to deal with this case. To avoid unnecessary technicalities, we discuss only that due to Abel-Poisson.

DEFINITION 3.2. Consider the series $\Sigma_{n=-\infty}^{\infty} c_n r^n e^{inx}$ for $0 \leq r < 1$. If this series converges to $S(r,x)$ and $\lim_{r \to 1^-} S(r,x) = s(x)$ exists, we say that (3.4) is Poisson summable to $s(x)$.

For example, the series $\Sigma_{n=1}^{\infty} \cos nx$ is Poisson summable to $-\frac{1}{2}$ as long as $x \neq 0 \bmod 2\pi$.

Since the convergence factor $r^n = e^{n \ln r}$ decreases faster than any power of n for large values of $|n|$, we expect Poisson summability to serve for any finite order derivative of a Poisson summable Fourier series. Although this allows us to proceed in an operational sense, it leaves unanswered the important question: what sort of mathematical objects are being represented by a Poisson summable Fourier series?

Our example above indicates that it might be natural to find the answer within the framework of generalized functions. For the series $\Sigma_{n=1}^{\infty} \cos nx$ is an attempt to represent a function which is $-1/2$ everywhere except at $x = 2n\pi$, where it is a δ-distribution. Once the convergence of a Fourier series is treated as convergence in the sense of generalized functions, a certain formal simplicity is gained. An important and useful facility is that term by term differentiation be well-defined. The drawback, of course, is that (3.4) now represents a distribution and need not be defined pointwise.

3.2 CONVERGENCE OF FOURIER SERIES

Before stating the general results, let us look at a simple case. Consider the complex form (3.4). Generally this series may not be convergent, but by integrating with respect to x, we obtain a series

$$F(x) = c_o\, x + \sum_{n \neq 0} c_n\, e^{inx}/(in) \; .$$

This series has an extra power of n in the denominator, and therefore, if c_n has the right behavior for large $|n|$, will converge in the classical sense for a series of continuous functions (see for example, Lemma 1.19). Let us suppose this is the situation. Then (3.5) converges also in the sense of $\mathcal{S}'$, and by the corollary to Lemma 2.10, it may be differentiated term by term with the resulting series again converging as a tempered distribution. As each term in this series is C^∞, the ordinary and distributional derivatives then coincide. Thus (3.4) equals $DF(x)$. The general result now follows.

THEOREM 3.3. Consider the series $\sum_{n=-\infty}^{\infty} c_n e^{inx}$, where c_n is defined for integer n and $c_n = O(|n|^k)$ for some integer k. Then the Fourier series converges to a unique tempered distribution ϕ.

Remark. We say that $f = O(g)$ if $|f/g|$ is bounded uniformly.

Proof. It should be noticed that the series defines a generalized function which is periodic with period 2π. Hence, it does not have compact support

If $k \le -2$, (3.4) converges uniformly to a continuous function which is bounded by $\Sigma|c_n| < \infty$ and a fortiori to a tempered distribution. For other values of k, let us repeat the trick discussed above and examine the series

$$c_o\, x^{k+2}/(k+2)! + \sum_{n\neq 0} c_n\, e^{inx}/(in)^{k+2} . \tag{3.5}$$

By assumption $|c_n / n^{k+2}| = O(|n|^{-2})$, and (3.5) converges uniformly to the continuous function $F_{k+2}(x)$. Further F_{k+2} is also the $\mathcal{S}'$

sum of (3.5). Once again, applying Lemma 2.10,

$$\phi = D^{k+2} F_{k+2} = \sum c_n e^{inx}$$

converges to a tempered distribution.

This result is surprisingly simple and sufficiently general to cover a wide variety of applications. In the examples below, we shall see how to interpret the growth conditions on the coefficients c_n in a natural way.

The converse of Theorem 3.3 is also true, but will not be proved here, [15]. It should be pointed out that some attention must be paid to defining the Fourier coefficients associated with a tempered distribution. We cannot use (3.4) immediately since e^{inx} is not a test function in $\mathcal{S}$. If, however, $\phi \in C^{\infty\prime}$ with support in $[0,2\pi]$, such c_n are well-defined as $e^{inx} \in C^{\infty}$. As is to be expected, the periodicity of ϕ will allow us to define its Fourier coefficients by the simple expedient of defining a distribution ϕ_c with compact support by $\phi_c(f) = \phi(f)$ when $\operatorname{supp} f \subset [0,2\pi]$ and $\phi_c(f) = 0$ when $\operatorname{supp} f \cap [0,2\pi] = \emptyset$. ϕ_c will be said to be associated with ϕ. In accordance with these remarks, the Fourier coefficients of ϕ are defined as

$$c_n = 1/2\pi \, \phi_c(e^{-inx}) \quad . \tag{3.6}$$

The claim is that these are bounded by a polynomial in n at large values and uniquely determine $\phi \in \mathcal{S}'$ by means of (3.4). To indicate how this might come about, consider

$$\phi(f) = \sum_{n=-\infty}^{\infty} \langle \phi_c , f(x+2\pi n) \rangle = \phi_c(g) \quad \text{for} \quad f \in \mathcal{S} \tag{3.7}$$

in which g is defined by

$$g(x) = \sum_{n=-\infty}^{\infty} f(x+2\pi n) \ . \tag{3.8}$$

From the estimate $|D^r f(x+2\pi n)| \le C(N)\ (1+n^2)^{-N}$, valid for any positive integers r and N with suitable $C(N)$, it is easy to see that g is a C^∞ function of period 2π . Thus, g may be expanded as

$$g(x) = \sum_{n=-\infty}^{\infty} d_n e^{inx} \ . \tag{3.9}$$

Moreover, $D^r g$ exists on $[0,2\pi]$ for any positive integer r and is again smooth so that $d_n = O(|n|^{-r})$ upon using (3.4) and integrating by parts. Substituting (3.9) in (3.7) and using the continuity of ϕ_c , we find

$$\phi(f) = \phi_c(g) = 2\pi \sum_{n=-\infty}^{\infty} d_n c_{-n} \tag{3.10}$$

with the latter series converging. The relation (3.10) shows that the coefficients $\{c_n\}$ for a tempered distribution determine a linear functional on the sequences $\{d_n\}$ of rapid decrease. It is then plausible that the c_n should have at most polynomial growth. Even though this fact is true, to prove it requires methods other than those at our disposal. Assuming $c_n = O(|n|^k)$ for some k , we may show that ϕ is uniquely determined by (3.4) with (3.6). Let $\psi = \Sigma_n c_n e^{inx}$ which is in $\mathcal{S}'$ by Theorem 3.3. Then for any $f \in \mathcal{S}$ regarding e^{inx} as an element of $\mathcal{S}'$, we have

$$\psi(f) = \sum c_n \langle e^{inx}, f\rangle = \sum c_n \langle (e^{inx})_c, g\rangle$$

$$= \sum_{n,m} c_n d_m \langle (e^{inx})_c, e^{imx}\rangle$$

$$= 2\pi \sum c_n d_{-n} .$$

Thus, $\langle \psi - \phi, f\rangle = 0$ for all $f \in \mathcal{S}$. In summary, we state the converse of Theorem 3.3.

THEOREM 3.4. Let $\phi \in \mathcal{S}'$ have period 2π. There exist coefficients $c_n = O(|n|^k)$ for some integer k such that (3.4) converges in $\mathcal{S}'$ to ϕ.

These last two theorems give necessary and sufficient conditions for (3.4) to represent a tempered distribution. The growth conditions on the c_n provide the temperateness and without these it is a much more delicate matter to spell out requirements which will place (3.4) in $\mathcal{D}'$. We will take up this point again in section 3.6.

3.3 EXAMPLES AND APPLICATIONS

As our first illustration, let us return to the function introduced in section 3.1 (see 3.3). The Fourier series has already been given as

$$x = \pi + \sum_{n \neq 0} i\, e^{inx}/n \qquad 0 \le x \le 2\pi .$$

One can write the original function on the whole real line as

$$f(x) = x - 2\pi \sum_{n \neq 0} H(x - 2\pi n) \quad . \tag{3.11}$$

Differentiation in the sense of $\mathcal{S}'$, together with example 2.4 (a), leads to

$$-2 \sum_{n=1}^{\infty} \cos nx = 1 - 2\pi \sum_{n \neq 0} \delta(x - 2\pi n) \quad .$$

This should be compared with our remarks in section 3.1 on Poisson summability. The series on the left-hand side does not converge in the usual sense, but as a distribution, it is a sum of δ-distributions at the jump discontinuities of f when defined for all values of x. A more compact form of the above relation is

$$\sum_{n=-\infty}^{\infty} e^{inx} = 2\pi \sum_{n=-\infty}^{\infty} \delta_{x-2\pi n} \quad . \tag{3.12}$$

Further differentiation of (3.12) gives

$$\sum_{n=-\infty}^{\infty} n^k e^{inx} = 2\pi(-1)^k \sum_{n=-\infty}^{\infty} \delta^{k+1}_{x-2\pi n} \tag{3.13}$$

wherein lies a useful observation.

Given coefficients c_n satisfying the hypothesis of Theorem (3.3), (3.15) permits us to associate the power behaviour with derivatives of δ-distributions. For example, if $c_n \sim n^k$ for large $|n|$, then (3.4) converges to ϕ which contains a δ^{k+1} singularity. Similarly, for the lower order behaviour, if c_n - (constant) $n^k \sim n^r$, (3.4) could have a δ^{r+1} singularity, and so on, until the $1/n^2$ behaviour is reached which provides a continuous function contribution to ϕ. We should be quick to note that although (3.13) might indicate that this behaviour is a property of the periodic jumps at

$2\pi n$, it is true for any jump of ϕ arising inside the interval $[0,2\pi]$ and then propagated by periodicity. This is seen clearly in the example below.

EXAMPLE (a) Consider a triangular "wave" with a pulse of magnitude h and duration a, repeating every 2π units of time x. The graph of such a function appears in Figure 3.2.

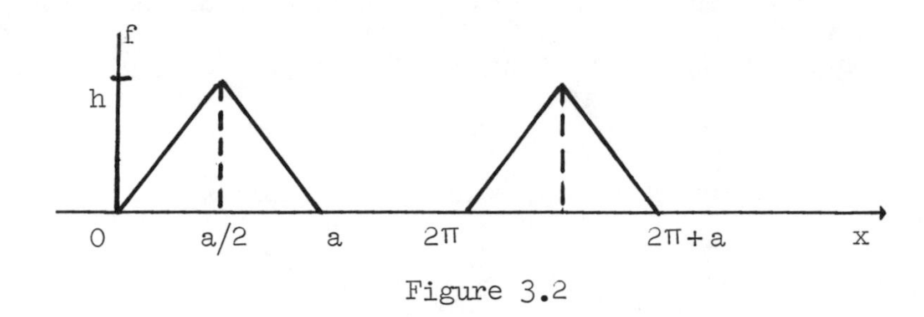

Figure 3.2

As a distribution, we may write $f(x) = \Sigma\, c_n\, e^{inx}$. The derivative of f has the expression

$$f' = 2h/a \qquad 2\pi n \le x \le 2\pi n + a/2$$
$$\quad -2h/a \qquad 2\pi n + a/2 < x \le 2\pi n + a$$

so that $f'' = 2h/a \sum_n \delta(x-2\pi n) + 2h/a \sum_n \delta(x-2\pi n - a) - 4h/a \sum_n \delta(x-2\pi n-a/2)$.

Using (3.13) we are able to rewrite this in the form

$$f'' = \sum_{n\neq 0} (in)^2\, c_n\, e^{inx} = h/(\pi a) \sum_n e^{inx}(1-e^{-ina/2})^2 \quad .$$

Therefore, for $n \neq 0$, $c_n = -(h/\pi a\, n^2)(1-e^{-ina/2})^2$. c_o can be computed directly as $c_o = ah / 4\pi$. Putting these pieces together results in the Fourier series for the triangular wave as

$$f(x) = ah/4\pi - h/\pi a \sum_{n\neq 0} e^{inx}(1 - e^{-ina/2})^2 / n^2 .$$

Taking derivatives once more, we have

$$Df = -ih/\pi a \sum_{n\neq 0} e^{inx}(1 - e^{-ina/2})^2 / n$$

which represents a series of step functions propagated with period 2π. A further derivative would produce a series of δ-distributions at the points $2\pi n$, $2\pi n + a/2$, and $2\pi n + a$ in accordance with the power behaviour $O(1)$.

PROBLEMS (1) Find the Fourier series development for the function e^{-x} on the interval $[0,2\pi]$. Does this series converge uniformly?

(2) Show that $S(x) = (\pi/2a) \cosh a(\pi - |x|) / \sinh a\pi$ on $[-\pi, \pi]$ has a Fourier series

$$S(x) = 1/2\, a^2 + \sum_{n=1}^{\infty} \cos nx / (n^2 + a^2)$$

and prove uniform convergence of this series. Use this result to deduce that $S(x)$ satisfies

$$D^2 S - a^2 S = -\pi \sum_{n=-\infty}^{\infty} \delta_{x-2\pi n} .$$

This is an example of a generalized differential equation which has a function as a solution. (<u>Hint</u>: Decompose $S(x)$ into two parts: for positive and negative values of x.)

(3) Use the expression for $S(x)$ in problem 2 to sum the Bernoulli series $\Sigma_{n=1}^{\infty} 1/n^2 = \pi^2/6$. (<u>Hint</u>: L'Hospital!)

(4) Let $f(x)$ have Fourier coefficients c_n. Then show

that $\sum_n c_n = \sum_n f(2\pi n)$ whenever this series converges.

(5) Show that $\sum_n e^{-4\pi^2 a n^2} = 1/2(\pi a)^{\frac{1}{2}} \sum_n e^{-n^2/4a}$.

(Hint: Consider the formula $\int dy\, \delta(x - y - 2\pi n) e^{-a y^2} = e^{-a(x - 2\pi n)^2}$ in conjunction with (3.12) .)

CHANGE OF INTERVAL. Suppose a function $f(x)$ is defined on the interval (a, b) . The Fourier series may be found by transforming to $(0, 2\pi)$ by setting $y = 2\pi(x - a)/(b - a)$ whereupon (3.12) becomes

$$\sum_n \exp[2\pi i\, n(x-a)/(b-a)] = (b-a) \sum_n \delta[x - a - n(b-a)] \quad .$$

The relations (3.4) are changed accordingly.

PROBLEM (6) Find a Fourier series for $f = \sin x$ defined on $0 < x < \pi$.

EVEN FUNCTIONS. Let $f(x)$ be defined on $[-\pi, \pi]$ such that $f(x) = f(-x)$. Then from (3.12)

$$\sum_{n=\infty}^{\infty} [\delta(x - \xi - 2\pi n) + \delta(x + \xi - 2\pi n)] = 1/\pi + 2/\pi \sum_{n=1}^{\infty} \cos nx \cos n\xi$$

and upon integrating formally with f from 0 to π

$$f(x) = 1/\pi \int_0^{\pi} d\xi\, f(\xi) + 2/\pi \sum_{n=1}^{\infty} \cos nx \int_0^{\pi} d\xi \cos n\xi\, f(\xi)$$

$$= \sum_{n=0}^{\infty} a_n \cos nx \quad .$$

This is a Fourier cosine series. Exactly the same result is true when f is an even generalized function ϕ where

DEFINITION 3.5. Denote $\dot{f}(x) = f(-x)$ for any test function in $\mathfrak{J}$. A unique generalized function $\dot{\phi} \in \mathfrak{J}'$ is defined by $\dot{\phi}(f) = \phi(\dot{f})$. ϕ is called even, if $\dot{\phi} = \phi$, and odd, if $\dot{\phi} = -\phi$.

ODD FUNCTIONS. Next we consider $f(x)$ defined on $(-\pi, \pi)$, but now $f(-x) = -f(x)$. As before, using (3.12) leads to an odd combination of δ-distributions

$$\sum_{n=-\infty}^{\infty} [\delta(x - \xi - 2\pi n) - \delta(x + \xi + 2\pi n)] = 2/\pi \sum_{n=1}^{\infty} \sin nx \sin n\xi \quad .$$

Integration with f from zero to π yields

$$f(x) = 2/\pi \sum_{n=1}^{\infty} \sin nx \int_0^{\pi} d\xi \, f(\xi) \sin n\xi$$

$$= \sum_{n=1}^{\infty} b_n \sin nx$$

which is a Fourier sine series. The same conclusion results for an odd generalized function.

PROBLEMS (7) Find a Fourier series for $\operatorname{sgn} x$ on $(-\pi, \pi)$. Use this to give the series for δ_x.

(8) Find a Fourier series for $|x|$ on $(-\pi, \pi)$. Using this series, verify problem 2.4 (4).

3.4 GIBBS PHENOMENON

Up to this point we have concerned ourselves with the distribution convergence of (3.4) and perhaps it is a good idea to remind ourselves how this contrasts with pointwise convergence. Our remarks following (3.13) would suggest that a Fourier series is a poor representation for a discontinuous function at its points of discontinuity. This may be seen in a well-known example shown in Figure 3.3; namely, the series for the step function $H(x-\pi)$ defined on $(0, 2\pi)$.

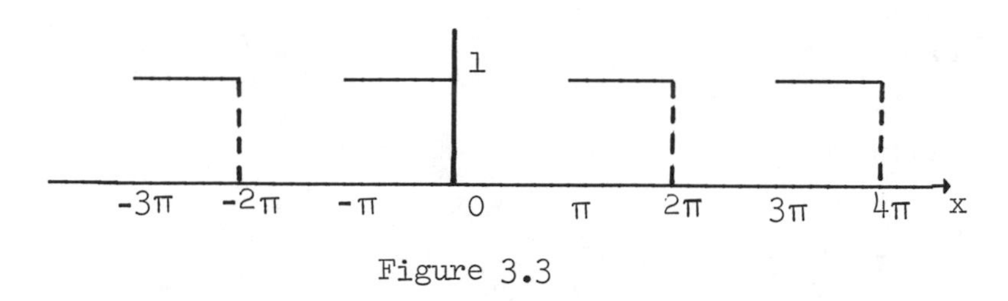

Figure 3.3

A direct calculation of the Fourier coefficients yields

$$H(x-\pi) = 1/2 - 2/\pi \sum_{n=0}^{\infty} \sin\,(2n+1)x\,/\,(2n+1)\ . \tag{3.14}$$

This series converges pointwise for all $0 < x < 2\pi$, but not uniformly in any neighborhood containing π. At this point, the series converges in the mean. Consider any $\varepsilon > 0$, and put $x = \pi \pm \varepsilon$. Then (3.14) gives

$$1/2 \pm 2/\pi \sum_{n=0}^{\infty} \sin\,(2n+1)\varepsilon\,/\,(2n+1)\ . \tag{3.15}$$

As expected, $H(x+\varepsilon-\pi) + H(x-\varepsilon-\pi) = 1$. Suppose we take a finite number of terms in (3.15) to represent the step function near $x = \pi$.

As we show below, for any $\epsilon > 0$, there exists an integer N such that

$$\sum_{n=0}^{N} \sin\,(2n+1)\epsilon\,/\,(2n+1) > \pi/4 \tag{3.16}$$

whereupon it is immediate that for this value of N , $1/2 \pm (2/\pi)\Sigma_{n=0}^{N} \sin\,(2n+1)\epsilon\,/\,(2n+1) \underset{<\,0}{>\,1}$. As N increases for decreasing ϵ , no matter how close we come to the discontinuity, a finite number of terms in (3.16) overshoots the value of the function to the right and undershoots it to the left. This type of behaviour is referred to as Gibbs' phenomena and is typical of a Fourier series at a point of discontinuity. It is interesting to ask whether there exist Fourier series which converge in the mean, but pointwise nowhere. For continuous functions the answer has recently been shown to be negative.

Let us give the proof of the inequality (3.16). Choose N to be the largest integer such that $(2N+1)\epsilon < \pi$ for a given $0 < \epsilon < \pi/(2n+1)$. Then, by area under the curve arguments

$$\sum_{n=0}^{N} \sin(2n+1)\epsilon\,/\,(2n+1) > \int_0^N dn\, \sin(2n+1)\epsilon\,/\,(2n+1)$$

$$= 1/2 \int_{\epsilon}^{(2N+1)\epsilon} d\alpha(\sin\alpha)\,/\,\alpha\ .$$

Consider the following integral decomposed into its positive and negative portions

$$\int_0^{\infty} d\alpha(\sin\alpha)/\alpha = \pi/2 = \int_0^{\pi} + \int_{\pi}^{\infty} d\alpha(\sin\alpha)/\alpha\ .$$

To check that the last integral on the right-hand side is negative, note

$$\int_{\pi}^{\infty} d\alpha\,(\sin\alpha)/\alpha = \sum_{n=0}^{\infty} \int_{(2n+1)\pi}^{(2n+3)\pi} d\alpha(\sin\alpha)/\alpha$$

$$= \sum_{n=0}^{\infty} \pi \int_{(2n+1)\pi}^{2(n+1)\pi} d\alpha(\sin\alpha)/\alpha(\alpha+\pi) < 0$$

from which we conclude $\int_0^{\pi} d\alpha\ (\sin\alpha)/\alpha > \pi/2$. An estimate for the expression above is a consequence of

$$\int_{\epsilon}^{(2N+1)\epsilon} d\alpha\,(\sin\alpha)/\alpha = \int_0^{\pi} - \int_0^{\epsilon} - \int_{(2N+1)\epsilon}^{\pi} d\alpha\ (\sin\alpha)/\alpha$$

$$> \int_0^{\pi} d\alpha\ (\sin\alpha)/\alpha - 3\epsilon \tag{3.17}$$

as the last two integrals are bounded by ϵ and 2ϵ respectively. Since ϵ may be made arbitrarily small, (3.17) now leads to the estimate

$$\sum_{n=0}^{N} \sin\,(2n+1)\epsilon/(2n+1) > 1/2 \int_0^{\infty} d\alpha\ (\sin\alpha)/\alpha = \pi/4 .$$

3.5 BESSEL'S INEQUALITY

Suppose f and g are continuously differentiable functions on $(0\,,2\pi)$. They then have uniformly convergent Fourier series

$$f = \sum_n c_n\, e^{inx} , \quad g = \sum_n d_n\, e^{inx} .$$

Inserting these into the integral below leads to

$$\int_0^{2\pi} dx\ \overline{f(x)}\ g(x) = 2\pi \sum_n \bar{c}_n\, d_n \tag{3.18}$$

with relevant interchanges of elementary infinite operations being permitted. This relation is called Parseval's theorem for Fourier series. It holds more generally for any Fourier series whose coefficients satisfy $\Sigma |c_n|^2 < \infty$.

More specifically, consider the Nth partial sum for one of these functions, $S_N(x) = \sum_{|n|\le N} c_n e^{inx}$ and $S_N(f) = \sum_{|n|\le N} |c_n|^2$. In (3.18), this gives

$$1/2\pi \int_0^{2\pi} dx\, |f(x)|^2 = S_N(f) + \sum_{|n|>N} |c_n|^2$$

from which follows the estimate $S_N(f) \le 1/2\pi \int_0^{2\pi} dx\, |f(x)|^2$. This last relation is referred to as Bessel's inequality for Fourier series and implies for $\sum_n |c_n|^2 < \infty$ convergence in the "mean" of the partial sums

$$\lim_{N\to\infty} \int_0^{2\pi} dx\, |f(x) - S_N(x)|^2 = 0 .$$

It is clear that we should not expect such results to hold for tempered distributions since Theorem 3.4 indicates that the Fourier coefficients for such a generalized function may have arbitrary polynomial growth and $\sum_n |c_n|^2$ need not exist. It is worthwhile to note that in the discussion of Theorem 3.4 we did make use of an analogue of Parseval's relation in which g was a tempered distribution and f in $\mathcal{S}$ (see (3.10)).

For functions f which are absolutely square-integrable on $(0, 2\pi)$, Parseval's relation provides an estimate on the error in approximating f in the "mean" by its partial sums. The "mean" error is just $\sum_{|n|>N} |c_n|^2$. In the same way, suppose $\phi \in \mathcal{S}'$ and $f \in \mathcal{S}$, and we approximate $\phi(f)$ in (3.10) by the partial sums up to order N. The error is similarly the remainder. This would seem to suggest a formal analogy between relations such as Bessel's inequality and Parseval's relation for absolutely square-integrable functions, and the linear functional approach to generalized functions. This is, in fact, the case and will be brought out further when we study the concept of duality in section 3.7.

3.6 FOURIER TRANSFORMS OF TEST FUNCTIONS

An essential part of the theory of generalized functions and its application rests on the concept of the Fourier transform. This is a natural extension to the infinite domain of Fourier's original ideas for finite intervals, or regions. For the definition, let f be absolutely integrable on the real line. Thus, we can define the Fourier transform of f, written as either $\tilde{f}$ or $\mathcal{F}(f)$, by the expression

$$\tilde{f}(p) = \int_{-\infty}^{\infty} dx\, e^{ipx} f(x) \quad . \tag{3.19}$$

This integral exists, since by assumption $|\tilde{f}(p)| \leq \int dx |f(x)| < \infty$, and defines an element of $C_o(R)$ (see Definition 1.16). If, moreover, $\tilde{f}(p)$ is absolutely integrable, the inverse Fourier transform is given by

$$f(x) = 1/2\pi \int_{-\infty}^{\infty} dp\, e^{-ipx} \tilde{f}(p) \quad . \tag{3.20}$$

In Chapters four and five, we shall need to treat the n-dimensional case for which (3.19) and (3.20) have the same form except as regards the notational changes appropriate to R^n .

$$p = (p_1, p_2, \dots, p_n) \qquad x = (x_1, x_2, \dots, x_n)$$

with

$$dx = dx_1 \, dx_2 \dots dx_n \qquad px = \sum_{i=1}^{n} p_i \, x_i \, . \tag{3.21}$$

An additional factor of $1/(2\pi)^n$ must be inserted into (3.20). A proof of the reciprocity between (3.19) and (3.20) under rather strong hypotheses is postponed until Proposition 3.6.

It is convenient to regard (3.19) as defining a mapping between functions $\mathfrak{F}: f \to \tilde{f}$ in which $\tilde{f} = \mathfrak{F}(f)$. The "inverse" map given by (3.20) will be written $\mathfrak{F}^{-1}$ accordingly. The properties of $\mathfrak{F}$ are the focus of our study.

To obtain the inversion formula (3.20), it is useful to consider the following contour integral in the complex plane of one variable p . The principal-value integral

$$\text{P.V.} \int_{-\infty}^{\infty} dp \, e^{ipx}/p = \lim_{\substack{R\to\infty \\ \epsilon\to 0^+}} \int_{-R}^{-\epsilon} + \int_{\epsilon}^{R} dp \, e^{ipx}/p$$

may be made up from a contour integral $\lim_{R\to\infty} \int_\gamma dp \, e^{ipx}/p$. Contours are shown in Figure 3.4. When $x > 0$, we complete γ in the upper half-plane with γ_+ and invoke Cauchy's residue theorem to give $\int_{\gamma + \cup \gamma} dp \, e^{ipx}/p = 0$. Along $\gamma_+, p = R \, e^{i\theta}$; consequently

$$\left|\int_{\gamma+} dp\, e^{ipx}/p\right| \leq \int_0^{\pi/2} d\theta(e^{-Rx\sin\theta} + e^{-Rx\cos\theta})$$

$$\leq \int_0^{\pi/2} d\theta(e^{-2Rx\,\theta/\pi} + e^{-Rx(\pi/2-\theta)})$$

$$= 1/Rx\;(\pi/2 + 1 - \pi e^{-Rx}/2 - e^{-\pi xR/2}) \quad x > 0$$

which tends to zero as $R \to \infty$ for all $x > 0$. In the second inequality, we have used the estimates, $\sin\theta \geq 2\theta/\pi$ and $\cos\theta \geq \pi/2 - \theta$, valid for $0 \leq \theta \leq \pi/2$. These are obtained from Figure 3.5 below.

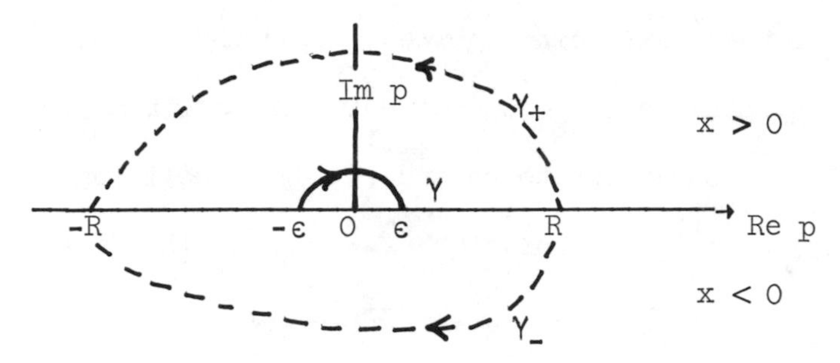

Figure 3.4

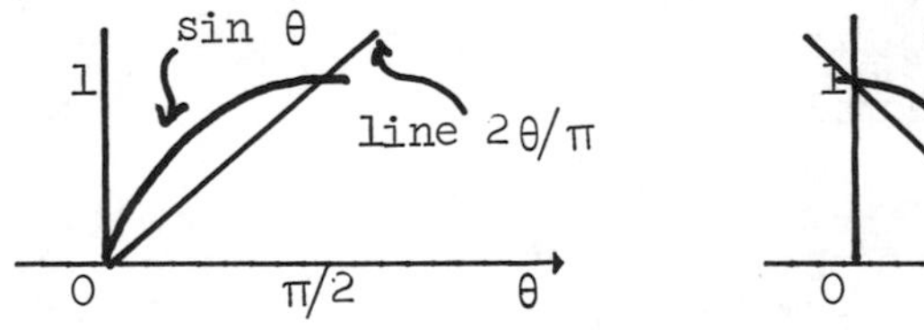

Figure 3.5

For the part of γ arising from the semi-circle about the origin, $p = \epsilon e^{i\theta}$, and $\int_{-\epsilon}^{\epsilon} dp\; e^{ipx}/p = i\int_{\pi}^{0} d\theta\,\exp(ix\epsilon e^{i\theta}) \to -\pi i$ as $\epsilon \to 0^+$ for all x. It then follows, that for $x > 0$, P.V.$\int dp\, e^{ipx}/p = \pi i$ A similar calculation for $x < 0$ results in a value $-\pi i$ for this integral. This is readily shown by completing γ to $\gamma \cup \gamma_-$ and taking into account the residue at $p = 0$ which is now enclosed by the contour. The final result is

$$\text{P.V.} \int_{-\infty}^{\infty} dp\, e^{ipx}/p = \pi i \operatorname{sgn} x \ . \tag{3.22}$$

From this, differentiation in the sense of generalized functions with example 2.4 (b), yields the representation

$$1/2\pi \int_{-\infty}^{\infty} dp\, e^{ipx} = \delta_x \ . \tag{3.23}$$

Recalling the definition of the δ-distribution in R^n (see 2.12), the n-dimensional analogue of (3.23) is

$$1/(2\pi)^n \int dp\, e^{ipx} = \delta_x \tag{3.24}$$

with the replacements indicated in (3.21). The relation (3.23) should be compared with its discrete analogue (3.12).

The last formula enables us to prove two useful results. The first is a weak form of the Fourier inversion formula (3.20) and the second is a continuous analogue of the Parseval relation (3.18).

PROPOSITION 3.6. Let $f \in \mathcal{S}$. Then $\mathfrak{F}[\mathfrak{F}(f)] = (2\pi)^n \dot{f}$.

Proof. A little later on, we will show that $f \in \mathcal{S}$ implies $\tilde{f} \in \mathcal{S}$ with respect to the p variables. Then

$$\mathfrak{F}[\tilde{f}](-y) = \int dp\, e^{-ipy} \int dx\, e^{ipx} f(x) = \lim_{N\to\infty} \int_{-N}^{N} dp\, e^{-ipy} \int dx\, e^{ipx} f(x)$$

$$= \lim_{N\to\infty} \int dx\, f(x) \int_{-N}^{N} dp\, e^{ip(x-y)} = \lim_{N\to\infty} (2\pi)^n (f * g_N)(y)$$

where $g_N = 1/(2\pi)^n \int_{-N}^{N} dp\, e^{ipx}$. However, by (3.24), $g_N \to \delta$ in $\mathcal{J}'$, and hence, by Proposition 2.26 and problem 2.7 (3)

$$\mathfrak{F}[\tilde{f}](-y) = (2\pi)^n \; f*\delta(y) = (2\pi)^n \; f(y)$$

which is the stated result.

COROLLARY. Under the hypothesis for Proposition 3.6, the inverse Fourier transform is given by (3.20).

It is possible to establish Proposition 3.6 under the weaker assumption that both f and $\tilde{f}$ are absolutely integrable. However, we shall not do so here.

PROPOSITION 3.7 (Parseval). Let f and $g \in \mathcal{S}$. Then we have

$$(2\pi)^n \int dx\; f(x)\; g(x) = \int dp\; \tilde{f}(-p)\; \tilde{g}(p)$$

Proof. Applying (3.19) directly, and making arguments similar to those in the previous proof, we find

$$\begin{aligned}
\lim_{N\to\infty} \int_{-N}^{N} dp\, \tilde{f}(-p)\, \tilde{g}(p) &= \lim_{N\to\infty} \int_{-N}^{N} dp \int dx\, e^{-ipx} f(x) \int dy\, e^{ipy} g(y) \\
&= \lim_{N\to\infty} \int dx\, dy\, f(x)\, g(y) \int_{-N}^{N} dp\, e^{-ip(x-y)} \\
&= (2\pi)^n \lim_{N\to\infty} \langle f\,, g*g_N \rangle = (2\pi)^n \langle f\,, g \rangle \\
&= (2\pi)^n \int dx\; f(x)\; g(x)\ .
\end{aligned}$$

The same result is true when f and g are absolutely square-integrable. In this case, Proposition 3.7 is the classical Parseval theorem for Fourier integrals.

PROBLEMS (1) Show that if f is absolutely integrable, then $\tilde{f} \in C_o(R^n)$.

(2) Show that $\mathfrak{F}[\delta(x-y)] = e^{ipy}$, then state the relevant nature of the distribution $\delta(x-y)$.

(3) Verify that $\mathfrak{F}[e^{bx}] = \delta(p-bi)$.

(4) Show that $\mathfrak{F}[x^m] = (-i)^m D^m \delta_p$.

We now turn to the main task of this section, which is to investigate the Fourier transforms of test functions. The first result reveals the characteristic feature of the space $\mathfrak{S}$.

THEOREM 3.8. The mapping $\mathfrak{F}: \mathfrak{S}(R^n) \to \mathfrak{S}(R^n)$, given by the Fourier transform, is a linear homeomorphism.

Remark. A homeomorphism between two topological spaces is a continuous bijective map with a continuous inverse.

Proof. Let $f \in \mathfrak{S}(R^n)$, then consider the expression

$$(1+\|p\|^2)^N D^r\tilde{f}(p) = (i)^{|r|} \int dx\, x^r f(x)(1-\Delta_n)^N e^{ipx} \qquad (3.25)$$

in which $\Delta_n = \Sigma_{i=1}^n \partial^2/\partial x_i^2$ is the n-dimensional Laplace operator. The rapid decrease of test functions in $\mathfrak{S}$ permits integration of (3.25) by parts with the discarding of the boundary terms. After this operation, (3.25) becomes

$$(1+\|p\|^2)^N D^r\tilde{f}(p) = (i)^{|r|} \int dx\, e^{ipx}(1-\Delta_n)^N x^r f(x) \qquad (3.26)$$

which is an absolutely convergent integral as the integrand is again a test function in $\mathfrak{S}$. A further estimate may be made on the right-hand side of (3.26). For an integer $M > n/2$, we have

$$\left|\int dx\, e^{ipx}(1+\|x\|^2)^{-M} (1+\|x\|^2)^M (1-\Delta_n)^N x^r f(x)\right|$$

$$\leq \sup_{x\in R^n} \left|(1+\|x\|^2)^M (1-\Delta_n)^N x^r f(x)\right| \int dx\, 1/(1+\|x\|^2)^M$$

$$\leq C(M)\ \|f\|_{2N}^{2M+|r|}$$

wherein $\|f\|_m^s$ is one of the $\mathfrak{S}$ norms (1.27) and $C(M)$ a suitable finite constant. When applied to (3.26), this estimate yields

$$\|\mathfrak{F}(f)\|_{|r|}^{2N} \leq C(M)\ \|f\|_{2N}^{2M+|r|} \quad . \tag{3.27}$$

It is worthwhile at this juncture to point out a certain "duality" between the power and differentiation indices for f , and its Fourier transform $\tilde{f}$ in this expression. We conclude from (3.27) that $\mathfrak{F}[f] \in \mathfrak{S}(R^n)$, and hence by the corollary to Proposition 3.6, $\mathfrak{F}^{-1}$ is given by (3.20). Thus, $\mathfrak{F}: \mathfrak{S} \to \mathfrak{S}$ is bijective and linear; that is, $\mathfrak{F}(\alpha f_1 + \beta f_2) = \alpha\mathfrak{F}(f_1) + \beta\mathfrak{F}(f_2)$ for complex numbers α and β , and for any functions f_1, f_2 for which (3.19) is defined.

The continuity of $\mathfrak{F}$ results immediately from (3.27). In considering a sequence $\{f_k\}$, tending to zero in the topology of $\mathfrak{S}$, (3.27) implies that $\{\mathfrak{F}[f_k]\}$ tends to zero in the topology of $\mathfrak{F}[\mathfrak{S}] \cong \mathfrak{S}$. Continuity for $\mathfrak{F}^{-1}$ is a consequence of the reverse estimate of (3.27) obtained by applying the same estimates on (3.20). This completes the proof of this theorem.

For the other class of test functions which we wish to investigate; namely $\mathcal{D}(R^n)$, Theorem 3.8 has the immediate consequence that $\mathfrak{F}[\mathcal{D}] \subset \mathcal{S}$. This is clearly a poor characterization of the Fourier transform of test functions in $\mathcal{D}$ as it desregards one of their essential features; specifically, their compact support. A sharper result can readily be obtained and is the content of Theorem 3.11 below. First, however, we shall discuss the one dimensional case.

Consider $f \in \mathcal{D}(R)$ with $\text{supp } f \subset [a, b]$, a finite interval. By Theorem 3.8, $\tilde{f}(p)$ exists and is C^∞ for all real values of p . Moreover, the key feature is that $\tilde{f}(p)$ is also defined for all complex values of p as may be seen from

$$\tilde{f}(p) = \int_a^b dx\, f(x)\, e^{i(\text{Re } p + i \text{ Im } p)} .$$

The same arguments as for real values of p show that $\tilde{f}(p)$ is C^∞ in both $\text{Re } p$ and $\text{Im } p$. Suppose γ is any smooth closed contour in the complex p plane, then

$$\begin{aligned} \int_\gamma dp\, \tilde{f}(p) &= \int_\gamma dp \int_a^b dx\, e^{ipx}\, f(x) \\ &= \int_a^b dx\, f(x) \int_\gamma dp\, e^{ipx} \\ &= 0 \end{aligned}$$

as $\int_\gamma dp\, e^{ipx} = 0$, e^{ipx} being analytic in p for all finite values of x . Thus, by Morera's theorem, $\tilde{f}(p)$ is analytic for all p in the finite complex plane; that is, $\tilde{f}(p)$ is an entire analytic function.

In order to completely characterize $\tilde{f}(p)$, it remains to ascertain the exponential growth in various directions. Again from Theorem 3.8, we know that $\tilde{f}(p)$ decreases faster than any power of

$|p|$ along the real axis. This same rate of decrease remains valid for any direction parallel to the real axis upon using the estimate

$$|p^r \tilde{f}(p)| = |(-i)^r \int dx\, f(x) \; d^r / dx^r \, e^{ipx}|$$

$$= |\int dx\, e^{ipx} f^{(r)}(x)| \le C(r, f) \max_{\substack{a \le x \le b \\ \mathrm{Im}\, p \in R}} [e^{-x\, \mathrm{Im}\, p}]$$

$$\le C(r, f)\, e^{c|\mathrm{Im}\, p|} \tag{3.28}$$

The constants are determined by $c = \max(|a|, |b|)$ and $C(r, f) = \int dx \max |f^{(r)}(x)|$. The bound (3.28) yields the exponential behaviour of $\tilde{f}(p)$ in directions parallel to the imaginary axis as exponential of type c determined by the "size" of the support of f.

Precisely the same result is true in the case of n dimensions. To state this concisely, we use a convenient characterization of entire functions in n variables satisfying estimates of the form (3.28) given by I. M. Gelfand and G. E. Shilov [2] . Reverting to multi-index notation, we make the following definition.

DEFINITION 3.9. Let $Z^k_m(C^n)$ denote the set of entire functions $f(z)$ on C^n such that for non-negative integers k and m

$$\| f \|^k_m = \sup_{\substack{|r| \le k \\ 0 \le c \le m \\ z \in C^n}} |z^r e^{-c|\mathrm{Im}\, z|} f(z)|$$

is finite.

To elaborate in more detail, $f \in Z^k_m$ means that $f(z)$ is an entire function of n complex variables $z = (z_1, z_2, \ldots, z_n)$ which satisfies the bound

$$|z_1^{r_1} z_2^{r_2} \cdots z_n^{r_n} f(z_1, z_2, \ldots, z_n)| \leq C(f,k,m)\exp[c(|\mathrm{Im}\, z_1| + \ldots + |\mathrm{Im}\, z_n|)]$$

for $0 \leq r_1 + r_2 + \ldots + r_n \leq k$ and $0 \leq c \leq m$. This is not quite a direct analogue of (3.28). Although $f(z)$ has exponential increase in a given variable, say z_j, along the imaginary directions, it has a fixed rate of decrease along the real directions rather than rapid decrease. The remedy for this is analogous to the construction of $\mathcal{S}^o$; namely, define $\mathcal{Z}_m(\mathbb{C}^n) = \bigcap_{k=0}^{\infty} \mathcal{Z}_m^k(\mathbb{C}^n)$. Functions $\mathcal{Z}_m$ now have rapid decrease in real directions as well as a fixed rate of increase in imaginary directions. The techniques expounded in section 1.7 may be applied to show that $\mathcal{Z}_m^k$ is a Banach space, and hence, by Theorem 1.12, $\mathcal{Z}_m$ is a complete countably normed space.

PROBLEMS (5) Show that $e^z = \prod_{i=1}^{n} e^{z_i}$ lies in $\mathcal{Z}_1^o$, but not in $\mathcal{Z}_m^k$ for any $k > 0$ and $0 \leq m < 1$. Provide an example of a function in $\mathcal{Z}_1^k$ for any positive integer k.

(6) Is $(\prod_{i=1}^{n} \cosh z_i)^{-1}$ a member of $\mathcal{Z}_1$?

(7) Verify that $\mathcal{Z}_m^k$ is a Banach space. (<u>Hint</u>: make use of the fact that a sequence of analytic functions is uniformly convergent on compact sets to a function which is analytic.)

A further definition will allow an arbitrary rate of exponential increase in the function $f(z)$.

DEFINITION 3.10. Let $Z(\mathbb{C}^n) = \bigcup_{m=0}^{\infty} Z_m(\mathbb{C}^n)$.

As Z_m is a complete countably normed space and the inclusions

$$Z_o \subset Z_1 \subset \dots \subset Z_m \subset Z_{m+1} \subset \dots$$

clearly hold, the space Z is a strict inductive limit of the family Z_m . The compatibility of the topologies for $Z_m \subset Z_{m+1}$ is immediate as Z_{m+1} allows a more rapid exponential increase than Z_m . If the norms in Definition 3.9 tend to zero for a sequence $\{f_s(z)\}$ with $0 \le c \le m$, this will also be true if $0 \le c \le m+1$ (see Definition 1.13).

Elements of Z are entire functions of n complex variables which in any given variable have fast decrease along directions parallel to the real axis, but exponential increase in imaginary directions. The rate of this increase is arbitrary provided that it is of exponential type. The significance of Z lies in the central result

THEOREM 3.11. The map $\mathfrak{F}\colon \mathcal{D}(R^n) \to Z(\mathbb{C}^n)$, given by the Fourier transform, is a linear homeomorphism.

Proof. The proof consists of two parts and is rather lengthy because of repeated estimates of the form (3.28). The first part, showing that $\mathfrak{F}[\mathcal{D}] \subset Z$, is similar to the one dimensional case already examined above. We repeat these arguments both for clarity and emphasis of the techniques involved. In particular, the proof of analyticity is slightly different in outward form.

Given $f \in \mathcal{D}(R^n)$, there exists an n dimensional box of the form $B(a, b) = \{x \in R^n \mid a_i \le x_i \le b_i , i=1,2,\ldots,n\}$ such that $\operatorname{supp} f \subset B(a, b, f)$. Hence, (3.19) becomes

$$\tilde{f}(p) = \int_{B(a,b,f)} dx\, f(x)\, e^{ipx}$$

in which $p \in \mathbb{C}^n$ and $px = \Sigma_{i=1}^{n} (\operatorname{Re} p_i + i \operatorname{Im} p_i) x_i$. Clearly, arguments similar to the one dimensional case show that $\tilde{f}(p) \in C^\infty(\mathbb{C}^n)$. To show more; namely, that $\tilde{f}(p)$ is an entire function of the complex variables $p_1, p_2, \ldots, p_n$, consider the power series expansion about the origin

$$\tilde{f}(p) = \sum_{|r|=0} p^r D^r \tilde{f}(p)|_{p=0} / r! \qquad (3.29)$$

The coefficients arc bounded by

$$|D^r \tilde{f}(p)|_{p=0}| \le \int_B dx |x^r f(x)| \le \prod_{i=1}^{n} c_i^{r_i} C(f) = c^r C(f) \qquad (3.30)$$

in which $C(f) = \int_B dx\, |f(x)|$ and $c_i = \max(|a_i|, |b_i|)$. Inserting this estimate in (3.29), produces $|\tilde{f}(p)| \le C(f) \Sigma_{|r|=0} |c\, p|^r / r! = C(f) \exp(c|p|)$. Thus, (3.29) converges absolutely for all finite p ; that is $\tilde{f}(p)$ is an entire analytic function.

The growth estimates for $\tilde{f}(p)$ are obtained from the relation

$$|p^r D^m \tilde{f}(p)| = |(i)^{|m|} (-i)^{|r|} \int dx\, x^m f(x)\, D^r e^{ipx}|$$

$$\le \int_{B(a,b,f)} dx\, e^{|x \operatorname{Im} p|} |D^r (x^m f(x))| . \qquad (3.31)$$

An integration by parts takes the first line into the second line.

With the notation of (3.30), (3.31) leads to the estimate $|p^r D^m \tilde{f}(p)| \le C(r, m, f) \exp(c|\mathrm{Im}\, p|)$ where the constant $C(r,m,f) = \int_B dx\, |D^r x^m f(x)|$. This last bound, on referring to Definition 3.9, indicates that $\tilde{f}(p) \in Z_m(\mathbb{C}^n)$ for all $m \ge \max_{1 \le i \le n} \{c_i\}$. Thus, we have shown that $\mathfrak{F}[\mathfrak{D}] \subset Z$.

To prove the converse, and at the same time verify that the map $\mathfrak{F}$ is bijective, we make use of the fact that $\mathfrak{F}[\mathfrak{D}] \subset \mathfrak{S}$ for real values of p. Hence the inverse Fourier transform is given by (3.20) where $\tilde{f}(p) \in Z$. The rapid decrease of $\tilde{f}(p)$ along the real axis implies $f(x) \in C^\infty(R^n)$. In fact, for any multi-index r

$$D^r f(x) = (-i)^{|r|}/(2\pi)^n \int dp\, e^{-ipx} p^r \tilde{f}(p) \qquad (3.32)$$

is absolutely convergent since the integrand is again in Z.

To conclude the proof that $D^r f(x)$ has compact support, we indulge in a little contour shifting in the expression (3.32). Let γ be any smooth contour parallel to the real axis, and in the upper half-planes $\mathrm{Im}\, p_i > 0$ if $x_i < 0$, but in the lower half-planes $\mathrm{Im}\, p_i < 0$ when $x_i > 0$ (see Figure 3.6). We then claim

$$D^r f(x) = (-i)^{|r|}/(2\pi)^n \int_\gamma dp\, e^{-ipx} p^r \tilde{f}(p) . \qquad (3.33)$$

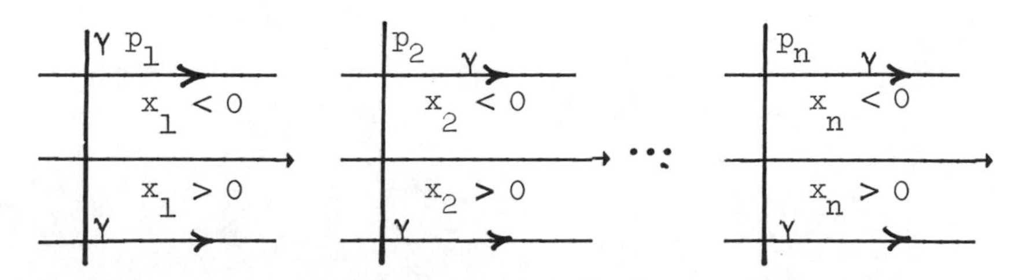

Figure 3.6

The verification of (3.33) rests upon Cauchy's theorem in n variables as applied to the contour $\gamma_1 \cup \gamma_R \cup \gamma_2 \cup \gamma_{-R}$ shown in Figure 3.7.

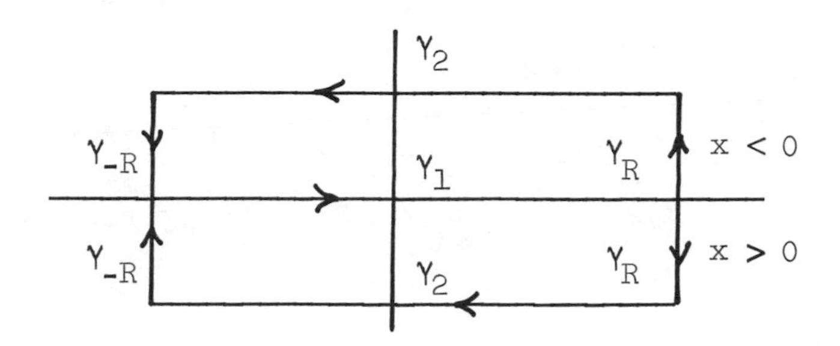

Figure 3.7

On the segments γ_R , γ_{-R} , we have typically an estimate

$$\left|\int_0^d d\mu\, e^{-iRx}\, e^{-|\mu x|}\, (R \pm i\mu)^r\, \tilde{f}(R \pm i\mu)\right|$$

$$\leq C(r,f) \int_0^d d\mu (e^{-|\mu x|} e^{c}) / (1 + R^2 + \mu^2) \qquad (3.34)$$

for any integer $N > 0$ and some constants c, C . This makes use of the fact that $p^r \tilde{f}(p) \in Z$. Letting $R \to \infty$, (3.34) tends to zero for any finite value of d ; consequently, $\lim_{R\to\infty} \int_{\gamma_R, \gamma_{-R}} dp\, p^r \tilde{f}(p) \cdot \exp(-ipx) = 0$. Finally, by Cauchy's theorem applied to the entire function, $e^{-ipx} p^r \tilde{f}(p)$,

$$\int_{\gamma_1 \cup \gamma_R \cup \gamma_2 \cup \gamma_{-R}} dp\, e^{-ipx}\, p^r\, \tilde{f}(p) = 0$$

so that in the limit as $R \to \infty$, $\int_{\gamma_2} \to \int_{\gamma_1}$, (3.33) follows.

Our last estimate is applied to (3.33) directly. On γ , set $p = \lambda + i\,\mathrm{Im}\,p$, and again bound $p^r \tilde{f}(p)$ by its characteristic estimate. Then from (3.33)

$$|D^r f(x)| \leq C(r,\tilde{f}) / (2\pi)^n \int d\lambda \exp(-|\mathrm{Im}\, p|\, [|x|-c])/(1+\lambda^2)^N$$

$$\leq M(r, N, \tilde{f}) \exp(-|\mathrm{Im}\, p|[|x|-c]) \qquad (3.35)$$

for any $N > 0$, some c, C. The constant $M(r, N, \tilde{f}) = C(r, \tilde{f})/(2\pi)^n \int d\lambda(1+\lambda^2)^{-N}$ is finite. We conclude that for $|x| > c$, (3.35) leads to $D^r f(x) = 0$ by taking $|\mathrm{Im}\, p|$ sufficiently large. $D^r f(x)$ then has compact support contained in the box $B(-c,c)$ as required.

Continuity of the map $\mathfrak{F}$ in both directions is more or less immediate from the estimates at our disposal. Suppose $\{f_s(x)\} \to 0$ in $\mathfrak{D}$, then also $C(r, m, f_s) \to 0$. Thus, $|p^r \exp(-|\mathrm{Im}\ p|c)D^m f_s(p)| \to 0$ since convergence in $\mathfrak{D}$ requires that c remain fixed. Continuity for $\mathfrak{F}^{-1}$ uses (3.35) with a similar observation for $M(r, N, \tilde{f})$. Thus, the proof of Theorem 3.11 is complete.

In spite of the foreshown lengthy demonstration, a number of useful results and techniques have appeared. Certainly each test function of compact support may be represented as the Fourier transform of a function in Z. Such is not the case for test functions from $\mathfrak{S}$ as $\tilde{f}$ need not be analytic (see problem 8 below). In addition, the topological nature of $\mathfrak{D}$ carries over to Z under the map $\mathfrak{F}$ and both are strict inductive limits of complete countably normed spaces. Theorem 3.11, as presented here, is a generalization of a result in analysis due to Paley and Wiener.

PROBLEMS (8) Treating $1/\cosh x$ as an element of (R), compute

$\mathfrak{F}(1/\cosh x)$. Verify that it is not a member of $Z(C)$. (Answer: $1/\cosh \frac{1}{2}p$).

(9) Obtain norm inequalities of the form (3.27) for the map $\mathfrak{F}: \mathcal{D} \to Z$. Show $Z^k(\mathbb{C}^n) = \bigcup_{m=0}^{\infty} Z^k_m(\mathbb{C}^n)$ is a strict inductive limit; next characterize the elements of $Z^k(\mathbb{C}^n)$. Finally, prove that $\mathfrak{F}: \mathcal{D}^{(k)}(R^n) \to Z^k(\mathbb{C}^n)$ is a linear homeomorphism.

3.7 FOURIER TRANSFORMS OF GENERALIZED FUNCTIONS

In attempting to give a meaningful definition of the Fourier transform for generalized functions, we meet a powerful concept much used in modern analysis: duality. This notion has had particular force in functional analysis and its manifold applications.

As ordinary functions are particular cases of generalized functions, one should seek a definition which reduces to the usual ones (3.19) and (3.20) whenever applicable. In this circumstance, the Parseval relation (Proposition 3.7) should hold. If we rewrite this, regarding g as a generalized function, then $\langle \tilde{g}, \dot{\tilde{f}} \rangle = (2\pi)^n \langle g, f \rangle$. Incorporating this formula directly, we define

DEFINITION 3.12. Consider $f \in \mathfrak{I}$ and $\phi \in \mathfrak{I}'$. The Fourier transform of ϕ, written $\tilde{\phi} = \mathfrak{F}(\phi)$, is the continuous linear functional on $\mathfrak{F}[\mathfrak{I}]$ as defined by the expression $\langle \tilde{\phi}, \dot{\tilde{f}} \rangle = (2\pi)^n \langle \phi, f \rangle$.

It is, of course, necessary to check that $\tilde{\phi}$ so defined is meaningful and has the announced properties. In this respect, strictly speaking, this should refer only to the case when $\mathfrak{I}$ is $\mathcal{D}$ or $\mathcal{S}$

as $\mathfrak{F}[C^\infty]$ remains to be defined!

The linearity of $\tilde{\phi}$ results from this same property for ϕ and the map $\mathfrak{F}$; that is,

$$\langle \tilde{\phi}, \alpha\dot{\tilde{f}}_1 + \beta\dot{\tilde{f}}_2 \rangle = \langle \tilde{\phi}, \widetilde{\alpha \dot{f}_1 + \beta f_2} \rangle = (2\pi)^n \langle \phi, \alpha f_1 + \beta f_2 \rangle$$

$$= (2\pi)^n \alpha \ \phi(f_1) + (2\pi)^n \beta \ \phi(f_2)$$

$$= \alpha \langle \tilde{\phi}, \dot{\tilde{f}}_1 \rangle + \beta \langle \tilde{\phi}, \dot{\tilde{f}}_2 \rangle \quad .$$

Similarly, the continuity of $\tilde{\phi}$ is a consequence of the same property for ϕ and $\mathfrak{F}$. We leave this step as an exercise. In this manner, we learn that the completeness of $\mathfrak{J}'$ implies that of $\mathfrak{F}[\mathfrak{J}']$ whenever this concept is satisfactorily defined by means of Definition 3.12.

PROBLEMS (1) Show that if $\{f_s\} \to 0$, then $\langle \tilde{\phi}, \dot{\tilde{f}}_s \rangle \to 0$ in Definition 3.12.

(2) Prove that $\mathfrak{F}[\mathfrak{J}']$, defined as in Definition 3.12, is complete when $\mathfrak{J}'$ is complete and $\mathfrak{F}$ is a linear homeomorphism.

The concept of duality is embodied in Definition 3.12. As we have seen, the relationship between the generalized functions and their test functions is such that $\mathfrak{J}'$ is the weak dual space to $\mathfrak{J}$. Theorems 3.8 and 3.11 have shown that the map $\mathfrak{F}: \mathfrak{J} \to \mathfrak{F}[\mathfrak{J}]$ is a linear homeomorphism; hence, $\mathfrak{F}[\mathfrak{J}]$ is itself a suitable space of test functions. The space $\mathfrak{F}[\mathfrak{J}']$ is then defined as the weak dual to $\mathfrak{F}[\mathfrak{J}]$ by Definition 3.12 and the relation between these duals is governed by this equation. A diagram is sometimes useful in remembering this concept, Figure 3.8.

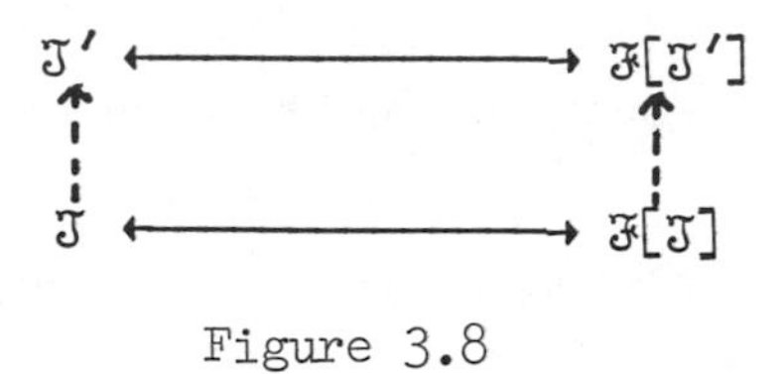

Figure 3.8

From Theorem 3.8, it follows that $\mathfrak{F}[\mathcal{S}'] \cong \mathcal{S}'$ whereas Theorem 3.11 gives $\mathfrak{F}[\mathcal{D}'] \cong Z'$, the continuous linear functionals on Z. The inverse relations are likewise valid by virtue of the duality these relations describe.

As a further illustration of the use of duality, we prove the extension of Proposition 3.6 to generalized functions.

THEOREM 3.13. For generalized functions ϕ, $\mathfrak{F}[\mathfrak{F}(\phi)] = (2\pi)^n \dot{\phi}$.

Proof. Recalling Definition 3.5, the following relations are easy to verify

$$\dot{\phi}(f) = \phi(\dot{f}) = 1/(2\pi)^n \langle \phi , \mathfrak{F}[\mathfrak{F}(f)]\rangle = 1/(2\pi)^n \tilde{\phi}(\tilde{f})$$

$$= 1/(2\pi)^{2n} \langle \mathfrak{F}[\mathfrak{F}(\phi)] , \dot{\mathfrak{F}}(\tilde{f})\rangle = 1/(2\pi)^n \langle \mathfrak{F}(\tilde{\phi}) , f\rangle .$$

PROBLEMS (3) Repeat problem 3.6 (3) with e^{bx} in $\mathcal{D}'$.

(4) Consider $\phi = \sum_{n=1}^{\infty} \delta(x-n)$ in $\mathcal{S}'$ or $\mathcal{D}'$. Find $\tilde{\phi}$ using Definition 3.12, and then show that this agrees with the formal result obtained from (3.19).

(5) Find the Fourier transform of the following $H(x)$, $\operatorname{sgn} x$, $D^r \delta_x$, $|x|$. Show that $\widetilde{D|x|} = \widetilde{\operatorname{sgn} x}$.

(6) Consider the sequence $\phi_n = (n/4\pi)^{\frac{1}{2}} e^{-nx^2/4}$ (example

2.4 (e)). Give an alternative proof of the fact that $\phi_n \to \delta_x$ by showing $\tilde{\phi}_n \to 1$ in $\mathcal{S}'$ and Z'.

(7) Repeat problem 6 for the sequence $\phi_n = \sin nx / x$.

As we have previously remarked, (3.19) cannot be used to define Fourier transforms of functions in C^∞. For example, x^m has a Fourier transform, $(-1)^{|m|} D^m \delta_p$, which is a generalized function. This might suggest that, for the duality of Definition 3.12 to be valid in this case, $\mathfrak{F}[C^{\infty\prime}]$ should be a test function space. Proceeding in this direction, recall in Theorem 2.20 that $\phi \in C^{\infty\prime}$ has compact support. As e^{ipx} is C^∞ in x, we may define the Fourier transform of ϕ as

$$\tilde{\phi}(p) = \langle \phi , e^{ipx} \rangle \quad . \tag{3.36}$$

Moreover, as $x^r e^{ipx}$ has the same property, $D^r \tilde{\phi}(p) = \langle x^r (i)^{|r|} \phi , e^{ipx} \rangle = (\widetilde{x^r (i)^{|r|} \phi})(p)$. Both of these relations are defined for p in $\mathbb{C}^n$.

It might now be expected that $\tilde{\phi}(p)$ is an entire analytic function. This is indeed the case as may be seen from the relations

$$\sum_{|r|=0} p^r D^r \tilde{\phi}(p)\big|_{p=0} / r! = \langle \sum_{|r|=0} p^r x^r (i)^{|r|} / r! \ \phi , 1 \rangle$$

$$= \langle e^{ipx} \phi , 1 \rangle = \langle \phi , e^{ipx} \rangle = \tilde{\phi}(p) \quad .$$

Thus, the power series on the left-hand side converges uniformly for all finite complex values of p. By reasoning similarly to the case of $\mathcal{D}$ in the last section, the compact support of ϕ will lead to an exponential type increase for $\tilde{\phi}(p)$ in directions parallel to the imaginary axes. To learn the difference between $\mathfrak{F}[C^{\infty\prime}]$ and Z, several examples will be helpful.

EXAMPLES (a) $\tilde{\delta}_x = \langle \delta_x , e^{ipx} \rangle = 1$

(b) For $\phi = H(x)H(1-x)$, $\tilde{\phi}(p) = (e^{ip} - 1)/(ip)$.

(c) $\widetilde{D^r \delta_x} = (-i)^{|r|} p^r$.

As the three examples above illustrate, $\tilde{\phi}(p)$ is an analytic function of exponential type, but not of rapid decrease in real directions. In fact, example (c) indicates that $\tilde{\phi}(p)$ may have increase in these directions. We are now ready to give a precise characterization for $\mathfrak{F}[C^{\infty\prime}]$.

DEFINITION 3.14. Let $\mathfrak{Y}_m^k(\mathbb{C}^n)$ denote the set of entire functions $f(z)$ of n complex variables, such that the norm

$$\| f \|_m^{1/k} = \sup_{\substack{0 \le r \le k \\ 0 \le c \le m \\ z \in \mathbb{C}^n}} | e^{-c|\mathrm{Im}\, z|} f(z) / (1 + |z|)^r |$$

is finite. Here k and m are positive integers.

It is left to the reader once more to show that $\mathfrak{Y}_m^k$ is a Banach space with the given norm. To define spaces with any polynomial increase, let us form

DEFINITION 3.15. Let $\mathfrak{Y}_m(\mathbb{C}^n) = \bigcup_{k=0}^{\infty} \mathfrak{Y}_m^k(\mathbb{C}^n)$, a complete countably normed space.

Provided that $m \geq 1$, each of the above examples are contained in $\mathcal{Y}_m$. The final space that we shall need to consider is the strict inductive limit obtained from the family $\{\mathcal{Y}_m\}$.

DEFINITION 3.16. Let $\mathcal{Y}(\mathbb{C}^n) = \bigcup_{m=0}^{\infty} \mathcal{Y}_m(\mathbb{C}^n)$ be the strict inductive limit of the spaces $\{\mathcal{Y}_m\}$.

From the last definition, we are finally able to give a precise statement on the nature of $\mathfrak{F}[C^{\infty\prime}]$.

THEOREM 3.17. The map $\mathfrak{F}\colon C^{\infty\prime}(R^n) \to \mathcal{Y}(\mathbb{C}^n)$ is a linear homeomorphism.

The detailed proof of this result is offered as problem 8 below. With Theorem 3.17, the duality relationship of Definition 3.12 enables us to determine the Fourier transforms of C^∞ functions as continuous linear functionals on the space $\mathcal{Y}$. The reader will not find it hard to provide examples for this last remark.

In conclusion, let us write down the inclusions between the various dual spaces of test functions and generalized functions that have been obtained so far. These are

$$\mathcal{D} \subset \mathcal{S} \subset C^\infty \Rightarrow Z \subset \mathcal{S} \subset \mathcal{Y}'$$

$$\mathcal{D} \subset C^{\infty\prime} \Rightarrow Z \subset \mathcal{Y}$$

$$C^\infty \subset \mathcal{D}' \Rightarrow \mathcal{Y}' \subset Z' .$$

PROBLEM (8) Give the proof of Theorem 3.17. (Hint: the various steps in the proof are similar to those in Theorem 3.11. To show

$\mathfrak{F}[C^{\infty\prime}] \subset \mathfrak{U}$, make use of Theorem 2.18 to prove that $\tilde{\phi}(p)$ increases in real directions no faster than $|z|^k$, where k is the order of ϕ . For the converse inclusion, consider $f(z) \in \mathfrak{U}$ and form $f(z)/(1+|z|^2)^N$, where N is chosen to make this expression integrable along real directions. This convergence factor is removed by differentiation.)

(9) Prove that $\mathfrak{U}$ is complete, and hence, give a proof that $C^{\infty\prime}$ is complete.

3.8 CONVOLUTIONS AND MULTIPLICATION

Convolution of test functions and generalized functions has been discussed in section 2.7, along with the relevant restrictions required by support constraints. In particular, for f and g in $\mathfrak{S}$, we found that $f*g$ was again in $\mathfrak{S}$, and by direct calculation $\widetilde{(f*g)}(p) = \tilde{f}(p)\,\tilde{g}(p)$. This relation suggests more. Namely, the right-hand side exists and is in $\mathfrak{S}$ for any $g \in \mathfrak{M}(\mathfrak{S})$; specifically, for $\tilde{g}$ the Fourier transform of a distribution in $C^{\infty\prime}$. Taking the inverse Fourier transform gives $f*g$ in $\mathfrak{S}$ for $g \in C^{\infty\prime}$, an alternative proof of Corollary 2 to Lemmas 2.24 and 2.25.

The difficulty in requiring the same relation to hold between generalized functions; viz, $\widetilde{\psi*\phi} = \tilde{\psi}\,\tilde{\phi}$, is quite clear. For sufficiently singular ψ and ϕ , this product need not be defined. The following proposition gives a result in this direction which covers the cases considered in section 2.7.

PROPOSITION 3.18. For $\phi \in \mathfrak{J}'$, let ψ be a generalized function such that $\psi * f \in \mathfrak{J}$ for any $f \in \mathfrak{J}$. If $\widetilde{\psi * f} = \tilde{\psi}\,\tilde{f}$ holds, $\widetilde{\phi * \psi}$ exists and is equal to $\tilde{\psi}\,\tilde{\phi}$.

Proof. From Definitions 2.23 and 3.5, we fine

$$\langle \phi * \psi \,, f \rangle = \langle \phi \otimes \psi \,, f(x+y) \rangle = \langle \phi \,;\, \langle \psi_y \,, f(x+y) \rangle \rangle$$
$$= \langle \phi \,,\, \dot{\psi} * f \rangle$$

Applying duality to both sides of this relation,

$$\langle \widetilde{\phi * \psi} \,,\, \dot{\tilde{f}} \rangle = \langle \tilde{\phi} \,,\, \widetilde{\dot{\psi} * f} \rangle = \langle \tilde{\phi} , \tilde{\psi}\,\dot{\tilde{f}} \rangle = \langle \tilde{\phi}\,\tilde{\psi} \,,\, \dot{\tilde{f}} \rangle$$

Unfortunately, this result is rather weak.

PROBLEM (1) Let g be a function in $\mathfrak{D}$ such that $\int dx\, g(x) = 1$. Define a sequence $\{g_n(x)\}$ in $\mathfrak{D}$ by $\tilde{g}_n(p) = \tilde{g}(p/n)$ $n = 1, 2, \ldots$. Show that $g_n \to \delta$ in $\mathfrak{D}'$.

4. INVARIANT DISTRIBUTIONS

A further application of the theory of generalized functions is considered relating to invariance with respect to a group of transformations on R^n . The notion of invariance is first defined for test functions and then extended to generalized functions by duality. We discuss three groups which are important for applications: translations, rotations, and Lorentz transformations.

4.1 GROUPS OF TRANSFORMATIONS

Throughout this chapter we shall be concerned with transformations of R^n onto itself. More precisely, such a transformation is a map $T: R^n \to R^n$ defined at all points, the image of x being $T(x)$. Various collections of such transformations admit algebraic properties with respect to the composition of successive transformations. Given maps $T_1, T_2: R^n \to R^n$, the composition of T_1 and T_2 written $T_1 \circ T_2$. is the map defined by

$$(T_1 \circ T_2)(x) = T_1[T_2(x)] \quad \text{all} \quad x \in R^n \quad . \tag{4.1}$$

The algebraic structure of interest is that of a group defined abstractly in

DEFINITION 4.1. A set G of elements $\{a,b,c,...\}$ is said to form a group with respect to a composition law $\circ$, if for any $a,b \in G$

there is an element of G, denoted $a \circ b$, called the composition of a and b for which

(i) given any $a,b,c \in G$, $a \circ (b \circ c) = a \circ b \circ c = (a \circ b) \circ c$

(ii) for any $a \in G$, there exists $e \in G$ such that $e \circ a = a$. The element e is independent of the choice of a.

(iii) for any $a \in G$, there exists $a^{-1} \in G$ such that $a^{-1} \circ a = e$.

A group G is called finite if it contains a finite number of distinct elements, otherwise it is called infinite. For a group G, the order of G is the number of distinct elements. In the following we shall omit the composition symbol between elements $a \circ b$, writing the latter as ab when it is clear that this product is composition in the group.

DEFINITION 4.2. Let G be a group such that $ab = ba$ for all elements $a,b \in G$. Then G is called commutative or abelian.

For an abelian group, we denote ab by $a+b$, the unit e by 0, and the inverse a^{-1} by $-a$.

EXAMPLES (a) The collection of positive, negative integers with zero forms an infinite abelian group with composition of two integers p and q defined as their sum $p+q$. The unit is 0, and the inverse of p is $-p$.

(b) The real numbers form an infinite abelian group under

addition. This is also true of R^n under the addition of n-dimensional vectors. Moreover, $R - \{0\}$ also forms an infinite abelian group with respect to multiplication of real numbers. In this case, the unit is the number 1 , while the inverse to the real number r is $1/r$.

(c) As an example of an abelian group of finite order n consider the set $\{1, \omega, \omega^2, \ldots, \omega^{n-1}\}$ in which $\omega = e^{2\pi i/n}$ is an n-th root of unity. Composition is the multiplication $\omega^p \omega^q = \omega^{p+q}$.

(d) Consider example (c) of section 1.1. There we defined $GL(n,C)$ as the set of $n \times n$ matrices with complex entries having non-zero determinant. This set forms a group in which matrix multiplication is the composition. The unit element is the $n \times n$ identity matrix, and requiring $\det A \neq 0$ for each $A \in GL(n,C)$ guarantees the existence and uniqueness of the inverse matrix A^{-1} . That $GL(n,C)$ is not commutative is easily settled by an example. Consider the matrices

$$A = \begin{pmatrix} 1 & 0 \\ 0 & -1 \end{pmatrix} \qquad B = \begin{pmatrix} 0 & -i \\ i & 0 \end{pmatrix}$$

then $AB \neq BA$. Further examples of matrix groups will appear in the main body of the chapter.

Before continuing we should settle one point concerning Definition 4.1. Strictly speaking (ii) and (iii) assume only the existence of a left unit and a left inverse. It is the case that these are also a right unit and inverse; therefore, unique. This is left as an exercise.

4.2 TRANSLATIONS

We define a translation of R^n by a real n-dimensional vector a as the homeomorphism of R^n onto itself given by

$$T_a(x) = x + a \qquad \text{all} \quad x \in R^n \ . \tag{4.2}$$

Translations may be applied successively with the result

$$(T_a T_b)(x) = T_a[T_b(x)] = T_a(x+b) = x+(a+b) = T_{a+b}(x) \text{ all } x \in R^n$$

or

$$T_a T_b = T_{a+b} = T_b T_a \ . \tag{4.3}$$

This gives the composition law for translations by the vectors b and a taken in succession. Clearly, the identity translation is translation by the vector 0, while the inverse to T_a is T_{-a}. Both of these properties are a direct consequence of (4.3) as well as the associative law (requirement (i) in Definition 4.1). Summarizing, we make the following definition.

DEFINITION 4.3. Let $T(n,R) = \{T_a, a \in R^n \mid T_a \text{ is a translation } (4.2)\}$ denote the infinite abelian group of translations on R^n.

Classes of functions which have special properties with respect to translations are of considerable use in applications. One important class is the class of translation invariant functions.

DEFINITION 4.4 Consider a test function $f \in \mathfrak{J}(R^n)$. f is said to be translation invariant if $f(x+a) = f(x)$ for all x, $a \in R^n$.

It follows immediately, that if f is translation invariant, $f(x) = f(0)$; and $Df = 0$ for derivatives in all directions. With a slight abuse of our notation, we define the translate of any function f by the vector a as

$$(T_a f)(x) = f(x+a) \quad . \tag{4.4}$$

Thus, for f translation invariant, $T_a f = f$ for all $a \in R^n$. This idea may be carried over to generalized functions by duality.

DEFINITION 4.5. For $\phi \in \mathfrak{J}'$, the translate of ϕ by a vector $a \in R^n$ is the generalized function $T_a \phi$ defined by $\langle T_a \phi , f \rangle = \langle \phi , T_{-a} f \rangle$ for all $f \in \mathfrak{J}$.

This is a well-defined expression since $f \in \mathfrak{J}$ certainly implies $T_a f \in \mathfrak{J}$. From this idea we may extend the concept of translation invariance to generalized functions in the following.

DEFINITION 4.6. A $\phi \in \mathfrak{J}'$ is said to be translation invariant if $T_a \phi = \phi$ for all $a \in R^n$.

Examples of these concepts are given in the problems.

PROBLEMS (1) Show that an alternative definition of derivative for a generalized function may be given by $D\phi = \lim_{a \to 0} (T_a \phi - \phi) / a$. The limit is a limit in the sense of generalized functions. Show that a translation invariant ϕ has zero derivative, and use this expression to give a definition of directional derivative for ϕ.

(2) Show that $T_a \delta_x = \delta_{x+a}$.

(3) Verify that for test functions f, $\widetilde{T_a f} = e^{-ipa} \tilde{f}$, and then show $\widetilde{T_a \phi} = e^{ipa} \tilde{\phi}$ for generalized functions.

(4) Consider an absolutely integrable function f. Show that $f*1$ is translation invariant and extend this result to $\phi*1$ where ϕ is a suitable generalized function.

Recalling our discussion at the beginning of section 2.7 concerning functions of several vector variables, consider $f(x,y)$ where both x and y lie in R^n. In other words f is defined on $R^n \times R^n \cong R^{2n}$. For such functions, translation invariance imposes strong restrictions on the functional relation between x and y. In fact,

THEOREM 4.7. Let $f(x,y)$ be a translation invariant function in $\mathcal{S}(R^n \times R^n)$. Then there exists a $g \in \mathcal{S}(R^n)$ such that $f(x,y) = g(x-y)$.

<u>Proof</u>. The translation invariance of f implies $f(x+a, y+a) = f(x,y)$ and, in particular, $f(x,y) = f(x-y, 0)$. As $f \in \mathcal{S}$ we may conclude that $g(x) = f(x,0)$ is the desired function. Clearly, the theorem applies equally well for any C^∞ function.

The next theorem extends this result to generalized functions for which further notation will be needed. Suppose $\phi(x,y) = \psi(x-y)$ are suitably defined functions regarded as generalized functions, namely,

$$\langle \phi , f \rangle = \int dx\, dy\ \phi(x,y)\ f(x,y) = \int dx\, dy\ \psi(x)\ f(x+y,y)$$
$$= \langle \psi , \check{f} \rangle$$

where

$$\check{f}(x) = \int dy\ f(x+y\ ,\ y)\ . \tag{4.5}$$

We will adopt this notation in the following.

THEOREM 4.8. Let $\phi \in \mathcal{D}'(R^n \times R^n)$ be translation invariant. Then there exists a $\psi \in \mathcal{D}'(R^n)$ such that $\langle \phi , f \rangle = \langle \psi , \check{f} \rangle$ for all $f \in \mathcal{D}(R^n \times R^n)$.

Proof. The hypothesis of translation invariance for ϕ leads to

$$\langle T_a \phi , f \rangle = \langle \phi , f \rangle = \langle \phi , T_{-a} f \rangle$$

or upon taking Fourier transforms

$$\langle \tilde{\phi} , \dot{\tilde{f}} \rangle = \langle \tilde{\phi} , e^{i(p+q)a}\ \dot{\tilde{f}} \rangle$$

for all $a \in R^n$ with $(p,q) \in R^{2n}$. Thus, $\langle [e^{i(p+q)a} - 1]\tilde{\phi} , \tilde{f} \rangle = 0$ for all test functions f . From the continuity of the map $\mathfrak{F}$, unless $\phi = 0$, it must be true that $e^{i(p+q)a} = 1$ for all vectors a . This can happen only if $p + q = 0$; and we deduce that $\mathrm{supp}\ \tilde{\phi} \subset \{(p,q) \in R^{2n} \mid p = -q\}$. Let $\tilde{\psi}$ denote the restriction of $\tilde{\phi}$ to this set. Then

$$\langle \tilde{\phi} , \dot{\tilde{f}} \rangle = \langle \tilde{\psi} , \dot{\tilde{f}}_R \rangle \tag{4.6}$$

in which $\tilde{f}_R(p) = \tilde{f}(p, -p)$. In terms of the Fourier transform this is

$$\tilde{f}_R(p) = \int dx\, dy\, e^{ipx - ipy} f(x, y) = \int dx\, dy\, e^{ipx} f(x+y, y)$$
$$= \tilde{\check{f}}(p)$$

and (4.6) is identical to the stated result. It is a straight forward matter to verify that $\check{f}$ is in $\mathcal{D}$ (or $\mathcal{S}$ if $f \in \mathcal{S}$).

PROBLEMS (5) Show that $T_a\, \delta_{x,y} = \delta_{x,y}$ for all $a \in R^n$ implies $\delta_{x,y} = \delta_{x-y}$ by direct calculation.

(6) Repeat Theorems 4.7 and 4.8 for the case of three vector variables, $f(x,y,z)$ on $R^n \times R^n \times R^n$, etc,

(7) Consider a translation invariant function $f \in \mathcal{D}$. Show that $f = 0$, and generalize this result to translation invariant $\phi \in C^{\infty\prime}$.

4.3 ROTATIONS

Consider the situation in which coordinates (x_1, x_2) in the plane undergo a rotation counterclockwise by an angle $0 \leq \theta < 2\pi$. With respect to new coordinates (x_1', x_2') (see Figure 4.1) the position of a point P in R^2 is given by the transformation equations

$$\begin{pmatrix} x_1' \\ x_2' \end{pmatrix} = \begin{pmatrix} \cos\theta & \sin\theta \\ -\sin\theta & \cos\theta \end{pmatrix} \begin{pmatrix} x_1 \\ x_2 \end{pmatrix}. \tag{4.7}$$

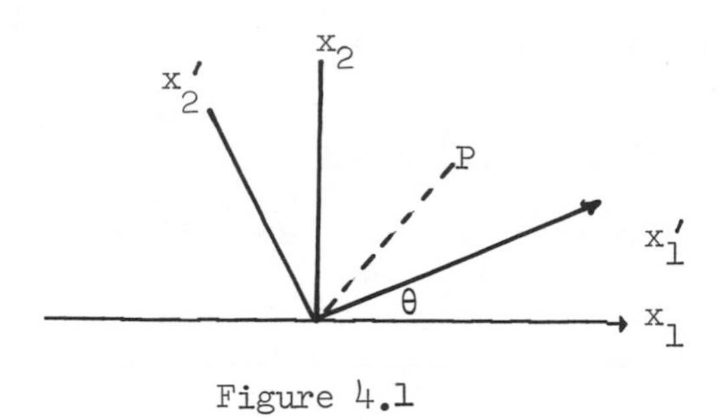

Figure 4.1

In order to generalize this to n dimensions, two features of (4.7) should be remarked; namely, both the lengths of vectors and the angle between two vectors remain unchanged.

In R^n, let x and y be any two vectors. The length of x is the usual Euclidean distance, $\|x\| = (x \cdot x)^{\frac{1}{2}}$. Similarly, the cosine of the angle between x and y is defined in analogy with the two and three dimensional cases by means of the scalar product, $\cos\theta\,(x,y) = x \cdot y / \|x\|\,\|y\|$. Consequently, preserving lengths and angles is equivalent to preserving the scalar product of two vectors. As an alternative notation for the scalar product, we shall employ

$$x \cdot y = (x\,,y) = \sum_{i=1}^{n} x_i\, y_i \tag{4.8}$$

and regard x as represented by an n component column vector, the entry in the i-th place being x_i. If A denotes an $n \times n$ matrix with entries a_{ij}, Ax is a column vector whose i-th entry is $\Sigma_{j=1}^{n} a_{ij}\, x_j$.

DEFINITION 4.9. A rotation in R^n is any real homogeneous transformation, $T\colon R^n \to R^n$ of the form $T(x) = x' = A\,x$, which leaves invariant the scalar product (4.8). The real matrix A will be called a rotation matrix.

It is worthwhile examining this last definition in greater detail. Given a rotation matrix A, consider $x' = Ax$ and $y' = Ay$. Preservation of their scalar product requires

$$(x', y') = (x, y) = (Ax, Ay) = (A^T A x, y)$$

where the transpose matrix of A, written A^T, is a matrix with entries $(A^T)_{ij} = a_{ji}$. This may be verified directly by use of (4.8). From these relations, we are led to

$$A^T A = 1 = A A^T \tag{4.9}$$

as the requirement for the matrix A to be a rotation. Such real matrices are called orthogonal and are, collectively, written $O(n)$, n indicating the number of dimensions. We also learn from (4.9) that one of the two possibilities, $\det A = \pm 1$, must hold. Orthogonal matrices with positive determinant are called proper, while those with negative determinant are termed improper. For the former, the customary notation is

$$SO(n) = \{n \times n \text{ real matrices } A \mid A^T A = 1 = A A^T, \det A = +1\} . \tag{4.10}$$

An alternative definition of a rotation may then be given using orthogonal matrices.

DEFINITION 4.9. A rotation in R^n is a mapping of the form $x' = A x$ in which $A \in O(n)$. Conversely, every such mapping defines a rotation. For a proper rotation, A must be an element of $SO(n)$.

PROBLEMS (1) Show that both $SO(n)$ and $O(n)$ form non-abelian groups.

(2) Show that every $A \in SO(2)$ has the form (4.7) for some angle θ. (<u>Hint</u>: Write $\begin{pmatrix} a & b \\ c & d \end{pmatrix}$ and use the constraints (4.9) to show that there exists a θ for which $a = \cos\theta$, $b = \sin\theta$, etc.)

(3) Let A be an improper rotation. Then prove that $A = I_s B$ where $B \in SO(n)$ and $I_s = -I$ for odd n, $\begin{pmatrix} -1_1 & & 0 \\ & \ddots & \\ 0 & & 1 \end{pmatrix}$ even n.

(4) Use (4.9) to prove that both the columns and rows of an orthogonal matrix form an orthonormal system of vectors for R^n.

Numerous useful and deep results stem from the concept of a rotation in R^n. Of these, we shall be concerned with functions which possess invariance with respect to proper rotations of the underlying coordinates. As for translations, the action of a rotation on a test function f is defined by

$$(A\,f)(x) = f(Ax) \qquad (4.11)$$

and for generalized functions by the duality

$$\langle A\,\phi\,, f\rangle = \langle \phi\,, A^{-1}f\rangle \qquad (4.12)$$

where $A \in O(n)$. It should be the case that $f \in \mathfrak{J}$ implies $Af \in \mathfrak{J}$; otherwise (4.12) will not define a generalized function

$A\phi \in \mathfrak{J}'$. This is an easy exercise. The relation (4.12) is motivated by again examining the classical situation in which ϕ is a function. There

$$\begin{aligned}\langle A\phi , f\rangle &= \int dx\ \phi(Ax)\ f(x) = \int d(A^{-1}x)\phi(x)\ f(A^{-1}x) \\ &= \int dx\ \phi(x)\ f(A^{-1}x) \\ &= \langle \phi , A^{-1}\ f\rangle\end{aligned}$$

where we have made use of the fact that if $x' = Ax$ is a proper rotation, the volume element transforms according to $dx' = \det[a_{ij}]dx = dx$. Once the action of rotations on functions and generalized functions has been given, the notion of invariance is immediate.

DEFINITION 4.10. A test function $f \in \mathfrak{J}$ is said to be rotation invariant if $Af = f$ for all $A \in SO(n)$. Similarly, $\phi \in \mathfrak{J}'$ is rotation invariant when $A\phi = \phi$.

In this last case, we have chosen to consider for convenience only proper rotations.

EXAMPLES (a) Any test function which is a function only of $\|x\|$ is rotation invariant, since by definition $(Af)(x) = f(Ax) = f(\|Ax\|) = f(\|x\|) = f(x)$. Typically in $\mathfrak{S}$ we have rotation invariant functions of the form $\exp(-\|x\|^2)$. As an example from $\mathfrak{D}$ we could take

$$\theta_n(x) = \begin{cases}\exp\left(-1/(1-\|x\|^2)\right) & \|x\| \leq 1 \\ 0 & \|x\| > 1\ .\end{cases} \tag{4.13}$$

This function is C^{∞} and has the unit ball for support.

(b) The δ_x-distribution is rotation invariant. For given any test function, we have $\langle A\delta_x , f\rangle = \langle \delta_x , f(A^{-1}x)\rangle = f(0) = \langle \delta_x, f\rangle$.

PROBLEMS (5) Show that the action of the rotation group commutes with the operation of taking the Fourier transform. Namely, for test functions prove that $\widetilde{(Af)}(p) = \tilde{f}(A\, p) = (A\,\tilde{f})(p)$ and use duality to extend it to generalized functions.

(6) Use the result of problem 5 to give an alternative verification that δ_x is rotation invariant.

(7) Give the proof that (4.12) is well-defined by demonstrating that the map in (4.11) is a homeomorphism of the test function spaces.

From these examples, one might suspect that rotation invariant functions are actually functions of $\|x\|$. This is, in fact, the case as we shall demonstrate below. Toward this end, it is convenient to introduce coordinates for R^n which are more natural to the rotation group. These are spherical coordinates which in n-dimensions take the form

$$x = r(\omega_1, \omega_2, \ldots, \omega_n) = r\,\hat{\omega} \quad r = \|x\| \quad \sum_{i=1}^{n} \omega_i^2 = 1 \qquad (4.14)$$

where $\hat{\omega}$ is a vector from the origin to the surface of the unit sphere. This latter surface we shall denote by $S^{n-1} = \{x \in R^n \,|\, \|x\| = 1\}$. If $d\Omega_n$ denotes the element of surface area on S^{n-1}, the volume element becomes

$$dx = r^{n-1}\, dr\, d\Omega_n \quad . \tag{4.15}$$

The surface area of S^{n-1} may be found in standard works on advanced calculus to be $\Omega_n = 2(\pi)^{n/2} / \Gamma(n/2)$, and for the volume of the unit ball, $B_n = \{x \in R^n |\ \|x\| \leq 1\}$ the value $m(B_n) = \Omega_n / n$.

EXAMPLE (c) As a further example of a rotation invariant function, consider $\phi = 1/r^{n-2}$. For test functions in $\mathbf{S}(R^n)$, $\phi(f) = \int dx\, f(x) / r^{n-2} = \int_0^\infty dr \int_{S^{n-1}} d\Omega_n\, r\, f(x)$ is a convergent integral. Thus, $\phi \in \mathbf{S}'$, but not in $C^{\infty\prime}$. This particular function arises in our discussion of the Laplace operator in section 5.2.

PROBLEM (8) The vector $\hat{\omega}$ in (4.14) has components ω_i , the direction cosines for R^n . Show that these may be uniquely specified by angles $0 \leq \theta_i \leq \pi$, $i = 1,2,\ldots,n-2$; $0 \leq \phi < 2\pi$ for which $\omega_1 = \cos\theta_1$, $\omega_2 = \sin\theta_1 \cos\theta_2$, $\omega_3 = \sin\theta_1 \sin\theta_2 \cos\theta_3$, ... , $\omega_{n-2} = \sin\theta_1 \ldots \sin\theta_{n-3} \cos\theta_{n-2}$, $\omega_{n-1} = \sin\theta_1 \ldots \sin\theta_{n-2} \cos\phi$, $\omega_n = \sin\theta_1 \ldots \sin\theta_{n-2} \sin\phi$. Also show that the surface element may be written $d\Omega_n = \sin^{n-2}\theta_1 \sin^{n-1}\theta_2 \ldots \sin\theta_{n-2}\, d\theta_1\, d\theta_2 \ldots d\theta_{n-2}\, d\phi$. Compare these with the usual expressions for $n = 2,3$.

With these kinematical preliminaries out of the way, the first idea which leads to the general form of a rotationally invariant function is that of averaging over the surface of the unit sphere.

DEFINITION 4.11. For $f \in \mathcal{J}(R^n)$, the spherical average of f is the function

$$f_S(r) = 1/\Omega_n \int_{S^{n-1}} d\Omega_n \, f(x) \ . \tag{4.16}$$

The function f_S is clearly a function only of the distance r of the point x from the origin and as such is invariant under rotations. As regards the topological properties of the spherical average, we may state the following.

LEMMA 4.12. Let $R_+ = \{r \in R \mid r \geq 0\}$. Then the spherical average in Definition 4.11 defines a map from $\mathfrak{J}(R^n)$ onto $\mathfrak{J}(R_+)$ which is linear and continuous in the sense that if $\{f_k\} \to 0$ then $\{f_{S,k}\} \to 0$.

Proof. The linearity of (4.16) is obvious, also the fact that $f \in C^\infty$ implies $f_S \in C^\infty(R_+)$. Moreover, if f has compact support so does f_S, and if f has rapid decrease so will f_S.

Let $\{f_k\}$ be a sequence tending to zero in one of the spaces $\mathfrak{J}$. Then from (4.16) we have

$$\sup_{0 \leq r < \infty} |r^q \, d^m/dr^m \, f_{S,k}| \leq \sup_{x \in R^n} |r^q \, d^m/dr^m \, f_k(x)|$$

from which the continuity follows directly in each of the cases $\mathcal{D}$, $\mathcal{S}$, or C^∞.

It is perhaps worth mentioning that (4.16) defines a map which is surjective but not injective. For example, a suitable odd function on the unit sphere will have spherical average zero. From Definition 4.11 we can deal directly with test functions and rotation invariance.

THEOREM 4.13. If $f \in \mathfrak{J}(R^n)$ is rotation invariant, then $f = f_S$.

<u>Proof</u>. Consider a proper rotation $x' = Ax$. The invariance of the volume element gives $dx' = dx$, while $r' = r$ implies from (4.15) that $d\Omega_n' = d\Omega_n$. It is also the property of the sphere S^{n-1} that any direction vector may be expressed in the form $\hat{\omega} = A\, e_n$ where A is a proper rotation and e_n is a unit vector in the direction of the n-th coordinate for R^n. Consequently, as f is rotation invariant $f(x) = f(r\,A\,e_n) = f(r\,e_n)$. Inserting this in (4.16) produces

$$\begin{aligned} f_S(r) &= 1/\Omega_n \int_{S^{n-1}} d\Omega_n\, f(r\,e_n) \\ &= f(r\,e_n) = f(x) . \end{aligned}$$

The extension of this result to generalized functions depends upon a suitable definition for the spherical average for ϕ. For this, consider again the classical situation of two functions f, g in which

$$\begin{aligned} \langle g_S , f \rangle &= \int dx\; g_S(r)\; f(x) = \int_0^\infty dr\; r^{n-1}\; g_S(r)\; f_S(r)\; \Omega_n \\ &= \int dy\; g(y)\; f_S(r) = \langle g , f_S \rangle . \end{aligned}$$

We carry this over directly to distributions.

DEFINITION 4.14. For $\phi \in \mathfrak{J}'(R^n)$, the spherical average of ϕ is the generalized function ϕ_S defined by the relation $\langle \phi_S , f \rangle = \langle \phi , f_S \rangle$ for all $f \in \mathfrak{J}(R^n)$.

The continuity of ϕ_S may be seen immediately since $f \to 0$ in $\mathfrak{J}(R^n)$

implies $f_S \to 0$ in $\mathfrak{J}(R^n)$ by means of Lemma 4.12. The other properties of ϕ_S are similarly verified. Our final result in this section is the analogue of Theorem 4.13 for distributions.

THEOREM 4.15. Let $\phi \in \mathfrak{J}'(R^n)$ be rotation invariant. Then $\phi = \phi_S$.

Proof. We make use of Theorem 2.28 and construct a regularization of ϕ by smooth functions. In particular, we choose the regularization $\phi_N = \phi * g_N$ in which

$$g_N(x) = \theta_n(x/N)(N/4\pi)^{n/2} \exp(-N \|x\|^2 / 4) \tag{4.17}$$

and $g_N(x) \to \delta_x$ by (2.30) and (4.13). Furthermore, each g_N is rotation invariant, and likewise the rotation invariance of ϕ implies that each ϕ_N is also rotation invariant. Thus, $g_N \in \mathfrak{D}$, and the first corollary to Lemma 2.24, gives $\phi_N(x)$ as a C^∞ rotation invariant function. Theorem 4.13 then shows $\phi_N = \phi_{N,S}$.

Assembling these remarks, and with the aid of the above definition, we are led to

$$\langle \phi , f \rangle = \lim_{N\to\infty} \langle \phi_N , f \rangle = \lim_{N\to\infty} \langle \phi_{N,S} , f \rangle = \lim_{N\to\infty} \langle \phi_N , f_S \rangle$$

$$= \langle \phi , f_S \rangle = \langle \phi_S , f \rangle$$

which implies that $\phi = \phi_S$. This is the statement for generalized functions that rotation invariance implies that ϕ is a "function of r alone" .

PROBLEMS (9) Prove that $\phi_N = \phi * g_N$ is rotation invariant if ϕ is rotation invariant and g_N is given by (4.17).

(10) Let $\hat{\omega}$ be any direction on the unit sphere S^{n-1}. Show that there exists a matrix $A \in SO(n)$ such that $\hat{\omega} = A e_n$, where $e_n = (0,0,\ldots,0,1)$. Use this result to show that any two points on S^{n-1} may be transformed into each other by means of a proper rotation. (Hint: Express A by means of the angles in problem 8.)

A particularly interesting case of this last result concerns distributions with support at the origin.

THEOREM 4.16. Let $\phi \in C^{\infty\prime}$ be rotation invariant with support at the origin. Then $\phi = P(\Delta_n)\delta_x$ where P is some polynomial and $\Delta_n = \Sigma_{i=1}^{n} \partial^2/\partial x_i^2$.

Proof. From section 2.6, ϕ must have the form

$$\phi = \sum_{|k|\leq m} c_k D^k \delta_x$$

for some integer m and constants c_k. By making use of problem 5, which states that ϕ is rotation invariant if, and only if, $\tilde{\phi}$ is also rotation invariant, and by Fourier transforming the above expression, this leads to

$$\tilde{\phi} = \sum_{|k|\leq m} c_k (i)^{|k|} p^k .$$

The required rotation invariance of this expression, together with the result of Theorem 4.13, implies that the only such polynomial is a function of $\|p\|^2$ alone. Thus, we may write $\tilde{\phi} = P(-\|p\|^2)$, whereupon $\phi = P(\Delta_n)\,\delta_x$ as stated.

4.4 LORENTZ TRANSFORMATIONS

As a last example of a transformation group with an interesting class of invariant distributions, we consider the Lorentz transformations. Our study will not be as complete as in the last two sections since difficulties of a rather technical nature present themselves and a full treatment would lead us too far from the objectives of this book.

It is convenient to single out a preferred co-ordinate x_o, representing the time in physical applications, and deal with vectors $x \in R^{n+1}$ where

$$x = (x_o, x_1, x_2, \ldots, x_n) = (x_o, \vec{x}) \quad . \tag{4.18}$$

The Euclidean distance in R^n (the space hyperplane) will continue to be written as $r = \|\vec{x}\|$. In order to define a Lorentz transformation, consider the following indefinite bilinear form

$$(x, gy) = x \cdot y = x_o y_o - \sum_{i=1}^{n} x_i\, y_i \tag{4.19}$$

in which we have again overburdened our use of the notation $x \cdot y$ and set g as a matrix with entries $g_{ij} = 0$, $i \neq j$; $g_{oo} = -g_{11} = \ldots = -g_{nn} = 1$. In accordance with the notation used in theoretical physics, we shall call $x \in R^{n+1}$

(i) time-like (future, past) when $x^2 > 0$ ($x_o > 0$, $x_o < 0$)

(ii) light-like (future, past) when $x^2 = 0$ ($x_o > 0$, $x_o < 0$)

(iii) space-like when $x^2 < 0$.

The various regions in which these conditions hold are respectively denoted by

(i) $V_\pm$ the future (past) cone

(ii) $L_\pm$ the future (past) light-cone

(iii) S the space-like region.

With respect to the form (4.19), R^{n+1} is divided into several disjoint regions

$$R^{n+1} = \{0\} \cup V_+ \cup V_- \cup L_+ \cup L_- \cup S \ .$$

These, schematically represented, appear as in Figure 4.2.

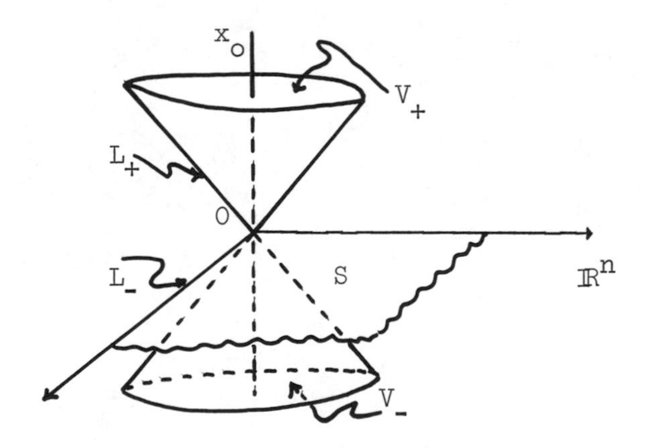

Figure 4.2

Following our general proceedure, we define the transformation group appropriate to the invariance of the bilinear form (4.19).

DEFINITION 4.17. A Lorentz transformation is a map $T: R^{n+1} \to R^{n+1}$

such that $T(x) = x' = \Lambda x$, where Λ is an $n+1 \times n+1$ real matrix which leaves invariant the form (4.19) .

In terms of the matrix g , the condition $(\Lambda x , g\Lambda y) = (x, g y)$ leads to the relation

$$\Lambda^T g \Lambda = g \tag{4.20}$$

which is the requirement that Λ should determine a Lorentz transformation. The collection of such matrices will be denoted by $\mathcal{L}_{n+1}$, the Lorentz group on R^{n+1} . As in the case of rotations, (4.20) implies $\det \Lambda = \pm 1$ with a corresponding distinction between proper and improper Lorentz transformations. Upon taking the oo-component of (4.20) and summing over repeated indices, we find $\Lambda^T_{oi} g_{ij} \Lambda_{jo} = g_{oo}$ or

$$\Lambda^2_{oo} = 1 + \sum_{i=1}^{n} \Lambda^2_{io} . \tag{4.21}$$

One of two possibilities occurs for Λ : $\operatorname{sgn} \Lambda_{oo} \gtrless 0$. These four possibilities lead to a decomposition of $\mathcal{L}_{n+1}$ into the disjoint sets

$$\mathcal{L} = \mathcal{L}^{\uparrow}_{+} \cup I_s \mathcal{L}^{\uparrow}_{+} \cup I_t \mathcal{L}^{\uparrow}_{+} \cup I_{st} \mathcal{L}^{\uparrow}_{+} . \tag{4.22}$$

We have denoted discrete transformations which perform reflections in the appropriate coordinate axes by

$I_t\, x = (-x_o , \vec{x})$ time inversion

$I_s\, x = (x_o , -\vec{x})$ n odd, $(x_o, -x_1, x_2, \ldots, x_n)$ n even: space inversion

$I_{st} x = I_s\, I_t\, x$ space-time inversion .

The set $\mathfrak{L}_+^\uparrow$ is given in our next definition.

DEFINITION 4.18. The set $\mathfrak{L}_+^\uparrow = \{\Lambda \in \mathfrak{L}_{n+1} \mid \det \Lambda = 1 , \Lambda_{oo} \geq 1\}$ consists of the proper, time-preserving Lorentz transformations.

In the decomposition (4.22), only $\mathfrak{L}_+^\uparrow$ is a group. This is a consequence of

LEMMA 4.19. For $\Lambda_1, \Lambda_2 \in \mathfrak{L}$, if sgn $\Lambda_{1,oo}$ and sgn $\Lambda_{2,oo}$ are both positive, then so is sgn $(\Lambda_1 \Lambda_2)_{oo}$.

Proof. Writing out the indicated matrix multiplications gives

$$(\Lambda_1\Lambda_2)_{oo} = \Lambda_{1,oo}\,\Lambda_{2,oo} + \sum_{i=1}^{n} \Lambda_{i,oi}\,\Lambda_{2,io}$$

$$\geq \Lambda_{1,oo}\Lambda_{2,oo} - \Big[\sum_{j=1}^{n} \Lambda_{1,oj}^2\Big]^{\frac{1}{2}}\Big[\sum_{k=1}^{n} \Lambda_{2,ko}^2\Big]^{\frac{1}{2}} \tag{4.23}$$

with the use of the Cauchy-Schwartz inequality on the last term. Employing the basic relation (4.20) $\Lambda = \Lambda g^2 = \Lambda g \Lambda^T g \Lambda$ and then multiplying this on both sides by $\Lambda^{-1} g$, we have

$$\Lambda g \Lambda^T = g \quad . \tag{4.24}$$

Taking the oo-component of this equation leads to the transpose of (4.21)

$$\Lambda_{oo}^2 = 1 + \sum_{i=1}^{n} \Lambda_{oi}^2 \quad . \tag{4.25}$$

Finally, substitution of (4.21) and (4.25) into (4.23) requires

$(\Lambda_1 \Lambda_2)_{00} \geq 1$. This concludes the proof.

A number of further kinematic properties of Lorentz transformations are to be found in the problems.

PROBLEMS (1) Verify that both $\mathcal{L}$ and $\mathcal{L}_+^\uparrow$ are matrix groups.

(2) Consider two dimensional Lorentz transformations in R^{1+1} . Show that there exists a number $-\infty < \alpha < \infty$ such that $\Lambda = \begin{pmatrix} \cosh\alpha & \sinh\alpha \\ \sinh\alpha & \cosh\alpha \end{pmatrix}$ where $\Lambda \in \mathcal{L}_+^\uparrow$.

(3) Show that $\Lambda \in \mathcal{L}_+^\uparrow$ if, and only if, $\Lambda x \in V_+$ for all $x \in V_+$. Hence, deduce that all the regions which form a disjoint partition of R^{n+1} with respect to (4.19) are left invariant by $\mathcal{L}_+^\uparrow$. Which are invariant with respect to $\mathcal{L}$?

(4) Choose any vector $x \in V_+$ and denote a unit positive time vector by $e_o = (1,\vec{0})$. Show the existence of a $\Lambda \in \mathcal{L}_+^\uparrow$ such that $x = (x^2)^{\frac{1}{2}} \Lambda e_o$. Use this result to show that $y \in V_-$ may be written $y = -(y^2)^{\frac{1}{2}} \Lambda e_o$ for some $\Lambda \in \mathcal{L}_+^\uparrow$. (Hint: First apply a rotation to bring x into the (x_o,x_n)-plane, and then treat the Lorentz part of the transformation as in problem 2.)

(5) Choose x in the space-like region S . Find a $\Lambda \in \mathcal{L}$ such that $x = (-x^2)^{\frac{1}{2}} \Lambda e_n$, where $e_n = (0,0,\ldots,0,1)$ is a unit space vector.

(6) Let e_o and e_n be the standard time and space unit vectors introduced in problems 4 and 5. Show that if $x \in L_+$, there is a $\Lambda \in \mathcal{L}_+^\uparrow$ for which $x = \Lambda(e_o + e_n)$.

(7) Show that $\Lambda \in \mathcal{L}_+^\uparrow$ if, and only if, $I_{st} \Lambda I_{st} \in \mathcal{L}_+^\uparrow$.

Turning our attention once again to test functions, we define the action of a Lorentz transformation by the usual relations

$$(\Lambda f)(x) = f(\Lambda x) \quad \langle \Lambda \phi , f \rangle = \langle \phi , \Lambda^{-1} f \rangle \text{ for all } f, \phi \ . \tag{4.26}$$

The reader may now easily verify that $f \in \mathfrak{J}$ implies $\Lambda f \in \mathfrak{J}$. The statement of Lorentz invariance of test functions or distributions is the content of the equation $\Lambda \phi = \phi$ for $\Lambda \in \mathfrak{L}_+^\uparrow$.

PROBLEM (8) Show that δ_x is Lorentz invariant.

When discussing the Lorentz invariance of a class of functions, it will be convenient as in the case of the rotation group, to introduce a set of coordinates which displays characteristic features appropriate to the Lorentz group. We will call such coordinates hyperbolic. First, introduce spherical coordinates for the space hyperplane R^n, that is,

$$x = (x_0 , r\hat{\omega}) \quad \|\hat{\omega}\| = 1 \quad dx = dx_0 \, r^{n-1} \, dr \, d\Omega_n \ . \tag{4.27}$$

Next, introduce a new variable s to account for the hyperbolic metric

$$s = x^2 = x_0^2 - r^2 \ . \tag{4.28}$$

The surfaces of constant s are the hyperboloids shown in Figure 4.3.

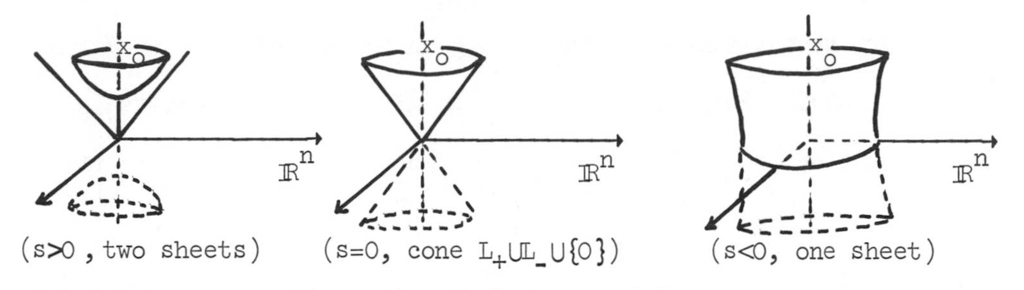

(s>0 , two sheets) (s=0, cone $L_+ \cup L_- \cup \{0\}$) (s<0, one sheet)

Figure 4.3

When expressed in terms of s , (4.27) becomes

$$\begin{aligned} &s>0 \quad x = (\pm[s+r^2]^{\frac{1}{2}}, r\hat{\omega}) \quad dx = \pm ds\; r^{n-1}\, dr\, d\Omega_n / 2(s+r^2)^{\frac{1}{2}} \\ &s<0 \quad x = (x_o, [x_o^2 + |s|]^{\frac{1}{2}}\hat{\omega}) \quad dx = -(x_o^2+|s|)^{(n-2)/2} dx_o\, ds\, d\Omega_n \quad . \end{aligned} \tag{4.29}$$

Furthermore, in the course of making some notational changes, it is convenient to modify the definition of the Fourier transform which appears in section 3.6. For integrable f , put

$$\tilde{f}(p) = \int dx\, e^{ipx} f(x) \qquad px = p_o x_o - \vec{p}\cdot\vec{x} \quad . \tag{4.30}$$

This amounts to changing the sign of the space part in $p = (p_o , \vec{p})$ in our previous expression, and has the advantage that

$$\widetilde{(\Lambda\phi)} = \Lambda\tilde{\phi} \quad . \tag{4.31}$$

Consequently, ϕ is Lorentz invariant, if and only if, $\tilde{\phi}$ is Lorentz invariant.

PROBLEM (9) Verify (4.31).

Consider a function f which is invariant under $\mathcal{L}_+^\uparrow$, and write

$$f = \tfrac{1}{2}(f + I_{st} f) + \tfrac{1}{2}(f - I_{st} f) = f_1 + f_2 \quad . \tag{4.32}$$

Problem 7 then implies, for arbitrary Λ with determinant +1, $\Lambda f_1 = f_1$ (even) and $\Lambda f_2 = \text{sgn}\, \Lambda_{oo}\, f_2$ (odd). Hence, (4.32) is a decomposition of f into even and odd parts with respect to time inversion. These are treated separately in the analogue of Theorem 4.13.

THEOREM 4.20. For $f \in C^o(R^{n+1})$, which is even and invariant under $\mathcal{L}_+^{\uparrow}$, there exists a $g \in C^o(R)$ such that $f(x) = g(x^2)$. When f is an odd function, it follows that $f(x) = \text{sgn}\, x_o\, g(x^2)$ and $g(x^2) = 0$ for space-like x.

Proof. We rely heavily on the results contained in problems 4, 5 and 6. Let us assume f is even. Then $f(x) = f(\Lambda e)$, where e is a standard vector determined as follows,

$$x = 0 \quad e = 0 \quad , \quad x^2 > 0 \quad e = (x^2)^{\frac{1}{2}}\, e_o \quad ,$$
$$x^2 = 0 \quad e = e_o + e_n \, , \quad x^2 < 0 \quad e = (-x^2)^{\frac{1}{2}}\, e_n \, .$$

Now define a function of x^2 by $g(x^2) = f(e)$. The continuity of g is clear since the invariance of g allows us to consider the difference $g(x^2) - g(y^2) = f(x) - f(y)$ as x^2 tends to y^2 along a sequence of values for which $|x_o - y_o| + \|x - y\|$ also tends to zero.

When f is odd, the same argument leads to $f(x) = \text{sgn}\, x_o g(x^2)$. To determine g for space-like x, consider the points $x = (k\, ,\, [|s| + k]^{\frac{1}{2}})$ and $I_t\, x$ on the hyperboloid $x^2 = s < 0$. Since f is odd $f(x) = -f(I_{st}x)$. Letting $k \to 0^+$ gives $f(e) = 0$ for $x^2 < 0$. The conclusion stated in the theorem now follows.

EXAMPLE. Test functions which are Lorentz invariant are less easily constructed than for the case of rotations. The reason is due to the fact that $\{0\}$ is the only compact set left invariant by all Lorentz transformations.

Consider $f(x) = \exp(-1/(x_0^2 - \|x\|^2))$ for $x^2 > 0$ and zero elsewhere. As a function of x^2, it is clearly Lorentz invariant and $\operatorname{supp} f \subset \overline{V_+ \cup V_-}$. Moreover, $f \in C^\infty$, but has neither compact support nor rapid decrease since $x_0^2 + \|x\|^2$ may become large while x^2 remains finite. Other examples of C^∞ Lorentz invariant functions are readily found by taking functions of the form $g(x^2)$. More generally, the observations contained in problems 4, 5 and 6 indicate that there are no non-trivial Lorentz invariant functions of rapid decrease or compact support on R^{n+1}.

PROBLEM (10) Show that there are no non-trivial members of $\mathcal{D}(R^{n+1})$ which are Lorentz invariant. Contrast the groups $O(n)$ and $\mathcal{L}_{n+1}$ in this respect.

Let us now turn our attention to Lorentz invariant distributions. From Theorem 4.20, we should expect some statement which is analogous in the singular case. Distributions with point support are easily handled.

THEOREM 4.21. For $\phi \in C^{\infty\prime}(R^{n+1})$ with support $\{0\}$, there exists a polynomial P such that $\phi = P(\partial^2/\partial x_0^2 - \Delta_n)\delta_x$.

<u>Proof</u>. The proof repeats that for Theorem 4.16 in all of its essen-

tial details. From (4.31), it is enough to examine the invariance of $\tilde{\phi}$ which is a polynomial in the p variables. By Theorem 4.20, this invariant polynomial is of the form $P(-p^2)$. The result follows upon taking Fourier transforms and noting that ϕ must be even.

Finally, let us approach the general question. If we mimic our earlier techniques for rotation invariance, we should introduce a hyperbolic average of the form

$$\bar{f}(s) = \int_{x^2 = s} dx\, f(x)$$

$$= \int dx\, \delta(x^2 - s)\, f(x) \qquad \text{for} \quad f \in \mathcal{S}(R^{n+1}) \ . \tag{4.33}$$

This last expression makes evident the invariance of $\bar{f}$. To investigate its support and smoothness properties, write out (4.33) in the two cases

$$\bar{f}(s) = \int_{R^n} r^{n-1}\, dr\, d\Omega_n \frac{f([s + r^2]^{\frac{1}{2}}, r\hat{\omega}) + f(-[s + r^2]^{\frac{1}{2}}, r\hat{\omega})}{2(s + r^2)^{\frac{1}{2}}} \tag{4.34}$$

when $s \geq 0$, and

$$f(s) = -\tfrac{1}{2} \int_{-\infty}^{\infty} dx_o \int_{S^{n-1}} d\Omega_n (x_o^2 + |s|)^{(n-2)/2} f(x_o\, , [x_o^2 + |s|]^{\frac{1}{2}}\hat{\omega})$$

when $s < 0$. It is evident from these formulas that f with compact support implies the same for $\bar{f}$ on R. Similarly, for $f \in \mathcal{S}$, the rapid decrease carries over to $\bar{f}$. Unfortunately, $\bar{f}$ being C^∞ does not! It is true, however, that only $s = 0$ presents any difficulty since f will be C^∞ on the set $R - \{0\}$. If we wish to retain the C^∞ property of generalized functions, along with the differentiation formula (2.15), then $\bar{f}$ should be regularized at the origin. Moreover, wishing to retain invariance properties of ϕ , this regularization should be carried out in a Lorentz invariant manner. This has been completed by P. Methée, [10], with a result analogous to Theorem 4.20.

We can see what is involved by looking at the singular behaviour of $f(s)$ near the origin. From (4.34)

$$\bar{f}(s) \sim C_1 \int_0^k dr\, r^{n-1} / (r^2 + s)^{\frac{1}{2}} \qquad s \geq 0$$

$$C_2 \int_0^k dx_o\, (x_o^2 + |s|)^{(n-2)/2} \qquad s < 0$$

in which $k > 0$, and C_1, C_2 are suitable constants. Then $\bar{f}(s) \sim C_3 \ln(1/|s|)$ for $n = 1$, while for $n = 2$, it behaves like a constant as $s \to 0$. For $n \geq 2$, the singular behaviour $\ln(1/|s|)$ occurs in a suitable derivative of $\bar{f}(s)$ near the origin. Further derivatives lead to power singularities. These algebraic singularities must be removed in a manner not unlike (2.3) in Chapter two. For further enlightenment, the interested reader should consult the work of Methée. Particular examples of Lorentz invariant distributions will abound in our discussion of the fundamental solutions for the wave operator appearing in sections 5.4 and 5.5.

5. PARTIAL DIFFERENTIAL OPERATORS

The methods of Fourier analysis and group invariance for generalized functions are brought together in the study of several partial differential operators from mathematical physics. Representations are obtained for their fundamental solutions and related Green's functions.

5.1 DIFFERENTIAL OPERATORS

A generalization of the differential operator, D^r, studied in section 2.4 may be given for polynomials of the form

$$P(x, D) = \sum_{|r| \leq N} a_r(x) D^r \tag{5.1}$$

in which the coefficients, $a_r(x)$, are C^∞ functions. If $a_r(x) \neq 0$ for at least one multi-index, r, such that $|r| = N$, $P(x,D)$ is said to be a linear differential operator of order N. For finite N, (5.1) defines a continuous linear map, $P(x,D)\colon \mathcal{J}' \to \mathcal{J}'$, by means of the relation

$$\langle P(x,D)\phi, f\rangle = \sum_{|r| \leq N} (-1)^{|r|} \langle \phi, D^r[a_r(x) f]\rangle . \tag{5.2}$$

The particular case in which the functions $a_r(x)$ are constant, say with values a_r, arises frequently in applications and permits a far reaching generalization of (5.1) by means of the Fourier transform. For this case, denoting (5.1) by $P(D)$, (5.2) simplifies to $P(D)\,\phi(f) = \phi(P(-D)f)$.

Consider a test function $f \in \mathcal{S}(R^n)$ and apply (3.25) to obtain the general relation

$$\widetilde{[P(D)\ f]}(p) = \sum_{|r|\leq N} (-i)^{|r|} a_r\, p^r\, \tilde{f}(p) = P(-ip)\ \tilde{f}(p) \ . \qquad (5.3)$$

Once this has been established for test functions, the dual relation of Definition 3.12 extends it to generalized functions. Indeed, for $\phi \in \mathcal{J}'$ and $f \in \mathcal{J}$, $\langle \widetilde{P(D)\phi}, \dot{\tilde{f}} \rangle = \langle \tilde{\phi}, \dot{\widetilde{P(-D)f}} \rangle = \langle \tilde{\phi}, P(-ip)\dot{\tilde{f}} \rangle$, whereupon

$$\widetilde{P(D)\ \phi} = P(-ip)\ \tilde{\phi} \ . \qquad (5.4)$$

When $\phi \in C^{\infty\prime}$, this relation results directly from (3.36). The merit in (5.4) lies in the observation that $\widetilde{P(D)}$ becomes a multiplication operator on $\mathfrak{F}[\mathcal{J}']$.

This connection between $P(D)$ and $P(-ip)$ is a particular example of a more general construction. Suppose $\tilde{u}$ is a function defined on C^n for which $\tilde{u}(-ip)\tilde{\phi} \in Z'$ when $\phi \in \mathcal{D}'$. By using the duality diagram (see Figure 5.1), we may define a generalized differential operator $u(D)$ which is a continuous linear map from $\mathcal{D}'$ into $\mathcal{D}'$. This statement is nothing more than the content of the equation

$$\langle \tilde{u}(-ip)\tilde{\phi}, \dot{\tilde{f}} \rangle = (2\pi)^n \langle u(D)\phi, f \rangle \quad \text{for} \quad f \in \mathcal{D} \qquad (5.5)$$

in which the left-hand side defines the distribution $u(D)\phi$. Another viewpoint

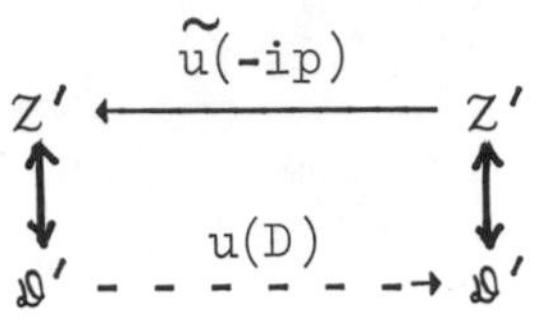

$$\begin{array}{ccc} Z' & \xleftarrow{\tilde{u}(-ip)} & Z' \\ \updownarrow & & \updownarrow \\ \mathcal{D}' & \overset{u(D)}{\dashrightarrow} & \mathcal{D}' \end{array}$$

Figure 5.1

on (5.5) is found by recalling the convolution discussed in section 3.8. Let u' denote that generalized function for which $u(-ip) = \mathfrak{F}(u')$. Then (5.5) gives meaning to the relation

$$u(D)\phi = u' * \phi \quad . \tag{5.6}$$

This idea, embodied in Figure 5.1 and (5.6), has been the subject of recent work related to the theory of linear partial differential operators having the form (5.1). Within this framework, $u(D)$ is an example of a pseudo-differential operator, or if we refer to the right-hand side of (5.6), a singular integral.

As we have noted, the operator $P(D)$ provides a linear continuous map from $\mathfrak{J}'$ into $\mathfrak{J}'$. It is interesting to ask under what conditions this map becomes a bijection. If it is surjective, onto, then for any $\psi \in \mathfrak{J}'$, there exists $\phi \in \mathfrak{J}'$ such that

$$P(D)\phi = \psi \quad . \tag{5.7}$$

This is the same question as asking if the differential equation (5.7) has a solution in $\mathfrak{J}'$ for an inhomogeneous term, ψ , also in $\mathfrak{J}'$. In practice, it is necessary to choose a more restrictive class for ψ in order to have an affirmative answer. When $\psi = \delta_x$, we rewrite (5.7) as

$$P(D)\gamma = \delta \tag{5.8}$$

and refer to γ as a fundamental solution for the operator $P(D)$. The map will be injective, (1-1) , when the relation

$$P(D)\phi = 0 \tag{5.9}$$

implies $\phi = 0$. In other words, the only solution to the homogeneous equation for (5.7) is trivial. Both of these questions, the existence and uniqueness of solutions to (5.7) with certain regularity properties, depends critically on the polynomial P and the domain in R^n within which we seek solutions. For second order operators, $N = 2$, we shall illustrate these features by examining three examples: the Laplace, heat and wave operators.

Before turning to these particular cases, we should point out that a particular solution to (5.7) is given by $\gamma * \psi$ provided the convolution exists. By adding the general solution, ϕ_0 , to (5.9), one obtains the general solution in the form $\phi_0 + \gamma * \psi$. The result of Lemma 2.24 would suggest that (5.7) is most amenable when $\psi \in C^{\infty\prime}$. This is, in fact, the case. However, with the exception of (5.8), we shall not attempt distribution solutions to (5.7). In addition, for the existence and uniqueness theorems governing (5.7), we refer the reader to the literature cited in the bibliography.

EXAMPLES (a) For the differential operator with constant coefficients, (5.6) becomes $P(D)\phi = u' * \phi$ where $u' = \Sigma_{|r| \leq N} a_r D^r \delta_x$.

(b) An early example of a singular integral which is not a differential polynomial was provided by

$$g(x) = \text{P.V.} \int_{-\infty}^{\infty} dy\, f(y) / (x - y) \quad . \tag{5.10}$$

Here, P.V. indicates the Cauchy principal-value defined as the limit $\lim_{\epsilon \to 0^+} \int_{-\infty}^{x-\epsilon} + \int_{x+\epsilon}^{\infty}$. In (5.10), g is called the Hilbert transform of f and, in this example, $u' = 1/x$ with $\tilde{u}(-ip) = i\,\pi\, \text{sgn}\, p$ (see (3.22)).

(c) The previous example generalizes to integral equations of the form

$$\phi(x) = \psi(x) + \int_{R^n} dy\, K(x, y)\, \phi(y) \tag{5.11}$$

in which the kernel, K , is translation invariant. Theorem 4.8 provides a k such that $K(x, y) = k(x - y)$ whereupon (5.11) assumes the form (5.6) with $u(D) = 1 - k*(\cdot)$.

As an application of the duality (5.5), let us treat (5.11) for the Cauchy kernel $K(x, y) = \lambda / (x - y)$ in (5.10). Such integral equations may be solved upon using the relation

$$1/x * 1/x = -\pi^2\, \delta_x \quad . \tag{5.12}$$

This is a consequence of the identity $\widetilde{(1/x * 1/x)}(p) = -\pi^2$. Rewriting (5.11) and convoluting with $1/x$ yields $\pi^2 \lambda \phi = 1/x * \psi$ $- 1/\lambda(\phi - \psi)$. Thus, we find

$$\phi = [\psi + \lambda(1/x * \psi)] / (1 + \lambda^2 \pi^2) \quad . \tag{5.13}$$

The a postiori justification for these computations is given for all ψ for which $1/x * \psi$ exists. The regularity of the solution ϕ is then determined by (5.13) .

PROBLEM. Consider (5.11) in one dimension. Show that a solution may be obtained by taking Fourier transforms. Identify zeroes of the denominator in the expression for $\tilde{\phi}$ with solutions of the homogeneous equation. Carry through this last part for the Cauchy kernel.

5.2 THE LAPLACE OPERATOR

Our first example is the n-dimensional analogue of Poisson's equation for the electrostatic potential ϕ in a region $V \subset R^n$ due to a charge density ψ

$$\Delta_n \phi = \psi \ . \tag{5.14}$$

The related homogeneous equation is Laplace's equation $\Delta_n \phi = 0$. The general theory of partial differential equations, [11], provides the correct boundary value problem for (5.14) which will lead to a unique solution ϕ with certain regularity conditions.

Consider a bounded, open, simply connected subset V of R^n with a boundary, ∂V, consisting of a finite number of portions each possessing a continuously varying tangent plane at interior points. The Dirichlet problem for V is to find a function ϕ, which satisfies (5.14) in V, and whose boundary value from directions inside V agrees with a given function α on ∂V. Replacing the boundary condition on ϕ by one for the outer normal derivative, $\partial\phi/\partial n$, which is required to be β, is called a Neumann problem for the region V. A mixed problem arises by specifying a real linear combination $\phi + \sigma\partial\phi/\partial n$ on ∂V. Finally, note that nothing is lost by restricting V to be simply connected since any connected region may be decomposed into simply connected components and each of the above boundary value problems applied to these.

The regularity assumptions on ∂V permit the application of Gauss' integral theorem on V to the vector field grad ϕ when ϕ is smooth enough; for example, $\phi \in C^2(V \cup \partial V)$. Our notation is such that grad $\phi = (\partial\phi/\partial x_1, \ldots, \partial\phi/\partial x_n)$ and div (grad ϕ) $= \Delta_n \phi$. If $\hat{n}$ denotes the unit outer normal to ∂V then $\partial/\partial n = (\hat{n}, \text{grad})$, whereupon Gauss' theorem reads

$$\int_V dx\, \mathrm{div}(\mathrm{grad}\ \phi) = \int_{\partial V} ds\, \partial\phi/\partial n\ . \tag{5.15}$$

The element of surface area on ∂V has been written as ds. For two functions, ξ and η, (5.15) produces Green's first identity

$$\int_V dx\, \xi \Delta_n \eta = -\int_V dx(\mathrm{grad}\ \xi, \mathrm{grad}\ \eta) + \int_{\partial V} ds\, \xi\, \partial\eta/\partial n\ . \tag{5.16}$$

Interchanging the roles of ξ and η in (5.16), we find Green's second identity

$$\int_V dx(\xi \Delta_n \eta - \eta \Delta_n \xi) = \int_{\partial V} ds(\xi\, \partial\eta/\partial n - \eta\, \partial\xi/\partial n)\ . \tag{5.17}$$

The first relation shows that a solution to the Dirichlet problem for V, if it exists, must be unique. Let ϕ_1 and ϕ_2 be two such solutions with the same boundary data. Then $\phi = \phi_1 - \phi_2$ is zero on ∂V and satisfies Laplace's equation inside V. Choosing $\xi = \phi = \eta$ in (5.16) yields $\int_V dx\, \|\mathrm{grad}\ \phi\|^2 = 0$, or ϕ is a constant on $V \cup \partial V$. From the boundary data, ϕ must be zero.

With regard to the Neumann problem for V, the source term and the boundary data must be compatible in the following manner. Suppose ϕ is a solution to (5.14), then (5.15) requires

$$\int_V dx\ \psi = \int_{\partial V} ds\ \beta \ .$$

If this condition holds, then ϕ is unique up to a constant. This results upon repeating the argument as in the Dirichlet case, only now the condition $\partial\phi/\partial n = 0$ does not fix the constant of integration.

It is appropriate at this point to ask whether solutions exist to either of the boundary value problems for (5.14). The answer is affirmative and we will content ourselves by indicating the role played by a fundamental solution toward this end.

For the Laplace operator, (5.8) assumes the form

$$\Delta_n \gamma_n = \delta_x \tag{5.18}$$

and defines a generalized function such that $\tilde{\gamma}_n(p) = (-1)/\|p\|^2$. This expression is integrable near the origin if $n \geq 3$, and we may consequently evaluate (3.20); that is,

$$\gamma_n(x) = (-1)/(2\pi)^n \int dp\ e^{ipx}/\|p\|^2 \ . \tag{5.19}$$

To exploit the rotation invariance of $\tilde{\gamma}$, introduce spherical coordinates with $\hat{p}\cdot\hat{x} = \cos\theta_1$ (see problem 4.3 (8)) to give

$$\begin{aligned} \gamma_n(x) &= -1/(2\pi)^n \int_0^\infty dr\ r^{n-3} \int_{S^{n-1}} d\Omega_n\ \exp(i\cdot\|p\|\ \|x\|\ \cos\theta_1) \\ &= \text{const.}/\|x\|^{n-2} \ . \end{aligned} \tag{5.20}$$

The constant in (5.20) is found by using (5.18) directly on a test function f which is a function of $r = \|x\|$ alone; namely, $f(0) = \langle\gamma_n , \Delta_n f\rangle = \text{const.}\ \Omega_n \int_0^\infty dr\ r\,[f''(r) + (n-1)/r\ f'(r)] = -\text{const.}\ (n-2)\Omega_n f(0)$

When $n=2$, (5.19) diverges logarithmically near $p=0$, suggesting that γ_2 behaves like $\ln(\|x\|)$ at large values of x . The result of a short calculation leads to

$$\gamma_n(x) = \begin{cases} -1/[(n-2)\Omega_n \|x\|^{n-2}] & n \geq 3 \\ (1/2\pi)\ln(\|x\|) & n = 2 \\ (1/2)\,[x\,H(x) - x\,H(-x)] & n = 1 \,. \end{cases} \tag{5.21}$$

A fundamental solution for the Laplace operator is then a C^∞ solution to Laplace's equation in the region $R^n - \{0\}$.

PROBLEMS (1) Let $A \in SO(n)$ and consider $x' = Ax$. Show that if $\gamma(x)$ is a solution to (5.18), then so is $\gamma(x')$. Hence γ is rotation invariant as may be seen explicitly in (5.21).

(2) Verify that the expressions for $n=1,2$ in (5.21) satisfy (5.18).

Let us introduce a second fundamental solution for the Laplace operator, by $\gamma = \gamma_n + \gamma_o$, where γ_o is chosen to satisfy the following boundary value problem

$$\Delta_n \gamma_o = 0 \text{ inside } V, \text{ with either } \begin{cases} \gamma = 0 \text{ on } \partial V \text{ (Dirichlet)} \\ \partial\gamma/\partial n = 1/m(\partial V) \text{ on } \partial V \text{ (Neumann)} \,. \end{cases} \tag{5.22}$$

The constant, $m(\partial V)$, in (5.22) is arbitrarily chosen to be the surface area of the boundary. When such a solution, γ_o , exists for the region V , γ is called a Green's function for the region. By means of γ , an integral representation for a solution to (5.14) can be obtained as follows. Suppose that $\phi \in C^2(V)$ and apply (5.17) in the y variables when $\xi = \phi(y)$, $\eta = \gamma(x-y)$ with x an arbitrary point in V. We then find

$$\phi(x) = \int_V dx\, \gamma(x-y)\psi(y) + \begin{cases} \int_{\partial V} ds\, \alpha(y)(\hat{n}, \mathrm{grad}_y\, \gamma(x-y)) & \text{(Dirichlet)} \\ -\int_{\partial V} ds\, \gamma(x-y)\beta(y) - \langle \phi \rangle_{\partial V} & \text{(Neumann)} \end{cases} \qquad (5.23)$$

in which $\langle \phi \rangle_{\partial V}$ is the average value of ϕ on the boundary. Although the relations (5.23) were derived by assuming that ϕ was a solution of the appropriate boundary value problem, it is still a valid question to ask whether (5.23) is that solution. It is clear that (5.23) satisfies (5.14) inside V from the properties of the fundamental solution. What must be demonstrated is that (5.23) has the correct boundary data. This will certainly be the case if α and β are in $C(\partial V)$ and $\psi \in C^1(V \cup \partial V)$, though weaker conditions do exist.

The problem of proving the existence of a solution ϕ has been reduced to solving the homogeneous problem (5.22). This depends only on the region V and the form of the differential operator. The existence of such Green's functions has been demonstrated for a large class of regions. However, an explicit form for γ is much harder to obtain. For regions with a high degree of symmetry such that the Laplace operator separates into single variable operators with respect to suitably chosen coordinates, γ may be obtained as a series of special functions. For numerical work, a small discontinuity in the boundary or boundary data can necessitate keeping a large number of terms in such series for an accurate representation of ϕ.

EXAMPLE. (Green's function for the sphere) For the unit sphere in R^n, the method of images from potential theory gives the Green's function for the Dirichlet problem with ease. If y is any point inside the sphere (centered at the origin), (5.21) gives the potential at x due to a charge $-1/(n-2)\Omega_n$ at y. Place a second charge of magnitude Q at a point y' outside the sphere to give a combined potential at x as $\gamma_n(x-y) + Q/(\|x-y'\|^{n-2})$. If we choose $y' = y/(\|y\|^2)$, the inverted image, and $Q = 1/[(n-2)\Omega_n\|y\|^{n-2}]$ then the required Green's function is

$$\gamma(x-y) = (-1/(n-2)\Omega_n)[1/\|x-y\|^{n-2} - 1/\|y\|^{n-2}\|x-y/\|y\|^2\|^{n-2})]. \quad (5.24)$$

PROBLEMS (3) Use the method of images to find the Dirichlet Green's function for the unit circle as $\gamma(x-y) =$
$(1/2\pi)\ \ln[\|x-y\| / (\|y\|\ \|x-y/\|y\|^2\|)]$.

(4) In problem 3, substitute $x = (r, \theta)$ and $y = (r', \theta')$ in the expression for γ. Then obtain Poisson's integral formula for the solution to the Dirichlet problem for Laplace's equation:

$$\phi(r,\theta) = (1/2\pi)\int_0^{2\pi} d\theta'\ \alpha(\theta')(1-r^2) / [1 - 2r\cos(\theta-\theta') + r^2].$$

5.3 THE HEAT OPERATOR

Let $\phi(t, x)$ denote the temperature at time t at a point $\vec{x}$ due to a heat source ψ in a heat conducting medium. In n-dimensions, ϕ is determined by the heat equation

$$(\partial/\partial t - \Delta_n)\phi = \psi. \quad (5.25)$$

As this relation is not invariant under time inversion, $t \leftrightarrow -t$, the solution to (5.25) will be defined for positive time. Bearing in mind this causal requirement, let γ be a fundamental solution; that is,

$$(\partial / \partial t - \Delta_n)\gamma = \delta_{(t,x)} \quad . \tag{5.26}$$

After taking Fourier transforms, we obtain an integral representation

$$\gamma(t,x) = 1/(2\pi)^{n+1} \int_{R^{n+1}} dp_o \, d\vec{p} \, e^{i(p_o t + \vec{p}\vec{x})}/(ip_o + \|\vec{p}\|^2) \quad . \tag{5.27}$$

By completing the contour in the upper (lower) half of the p_o complex plane according to $t > 0$ $(t < 0)$, Cauchy's residue theorem gives

$$\gamma(t\, , x) = H(t)/(2\pi)^n \int_{R^n} d\vec{p} \, e^{-(t\|\vec{p}\|^2 - i\vec{p}\cdot\vec{x})}$$

$$- H(t) \, (1/\pi \, t^{\frac{1}{2}})^n \prod_{j=1}^{n} I(x_j / t^{\frac{1}{2}}) \quad .$$

Here $I(u) = \int_0^\infty d\alpha \, e^{-\alpha^2} \cos(\alpha u) = (\pi/4)^{\frac{1}{2}} \exp(-u^2/4)$ as may easily be verified. Substituting this value for I in the above expression leads to

$$\gamma(t\, , x) = \gamma_t(x) = H(t)(1/4\pi t)^{n/2} \exp(-\|\vec{x}\|^2 / 4t) \quad . \tag{5.28}$$

One should note that for $t > 0$, $\gamma_t(x)$ is a solution to the homogeneous heat equation.

Consider the initial value problem for the homogeneous heat equation on R^n

$$(\partial / \partial t - \Delta_n)\phi = 0 \quad \text{with} \quad \phi(0\, , \vec{x}) = \alpha(\vec{x}) \quad t \geq 0 \, , \, \vec{x} \in R^n \quad . \tag{5.29}$$

By means of the fundamental solution derived above, the solution can be written in the form

$$\phi(t, \vec{x}) = \gamma_t * \alpha(\vec{x})$$

$$= H(t)/(4\pi t)^{n/2} \int_{R^n} d\vec{y} \exp(-\|\vec{x}-\vec{y}\|^2/4t)\alpha(\vec{y}) \quad . \qquad (5.30)$$

Suppose $\alpha \in \mathcal{S}'(R^n)$ and $t > 0$. Then $\gamma_t \in \mathcal{S}(R^n)$ and Corollary 1 to Lemma 2.24 indicates that $\phi(t, x)$ is C^∞ in all variables. Thus, any initial singularities are smoothed out at later times. The initial data, α , is recovered by taking the limit as $t \to 0^+$ in (5.30) and noting that $\gamma_t \to \delta_x$ with convergence in the $\mathcal{S}'$ topology. The expression (5.30) exhibits two additional features of the heat operator. Consider the case in which α has compact support. From (5.30), we see that at any point x for all positive t , the solution ϕ is influenced by all of the initial data. Physically, this effect is called diffusion. Moreover, this is true no matter how small the interval of elapsed time. Therefore, we say that the heat operator diffuses the initial data with infinite speed.

Let us close this section with some general remarks. First, by restricting the initial data in (5.29) to be tempered, the solution ϕ will be unique for all positive time t . This is not necessarily the case if $\alpha \in \mathcal{D}'$. Second, a solution to the inhomogeneous equation (5.25) may be obtained as $\gamma * \psi$, the convolution also being taken in the t variable as well as the space variables. Finally, solutions to a mixed initial value and Dirichlet (Neumann) problem for a bounded region $V \subset R^n$ may also be expressed in terms of a Green's function. They are, however, more complicated.

5.4 THE WAVE OPERATOR

In describing the time dependent behaviour of a wave field, $\phi(t, x)$, for a particle of mass m, it is necessary to solve the wave equation

$$(\partial^2/\partial t^2 - \Delta_n + m^2)\phi = \psi \ . \tag{5.31}$$

Such an equation arises frequently in applications; for example, quantum mechanics. The zero mass case arises in electromagnetic theory from Maxwell's equations and also in acoustics when studying the displacement of a vibrating membrane due to a source ψ. The particles of the wave field in these cases are, respectively, photons and phonons. The situation in which $m^2 < 0$ (imaginary mass) presents a definite mathematical problem, but one of marginal physical interest as a basic causal requirement for solutions to (5.31) is violated. Our discussion will deal with $m = 0$, but the analysis extends to the massive case with only minor modifications. These will be left as an exercise. In the following, it will be convenient to abbreviate the n-dimensional wave operator by the expression $\Box_n = \partial^2/\partial t^2 - \Delta_n$.

As previously, let us seek a fundamental solution and utilize it to solve an initial value problem. Our notation reverts to that of section 4.4 (see 4.18, 4.19 and 4.30) with the time variable as $t = x_o$. After Fourier transforming the relation

$$\Box_n \gamma = \delta \tag{5.32}$$

we consider the integral

$$\gamma(x) = -1/(2\pi)^{n+1} \int_C dp \; e^{i\,p\,x}/(p_o^2 - \|\vec{p}\|^2) \quad . \tag{5.33}$$

The contour C consists of integration over the space hyperplane, R^n , and a positive contour around the poles of the denominator in the p_o complex plane. Various contours are shown in Figure 5.2 and will be chosen to accomodate suitable boundary or initial conditions for the wave equation.

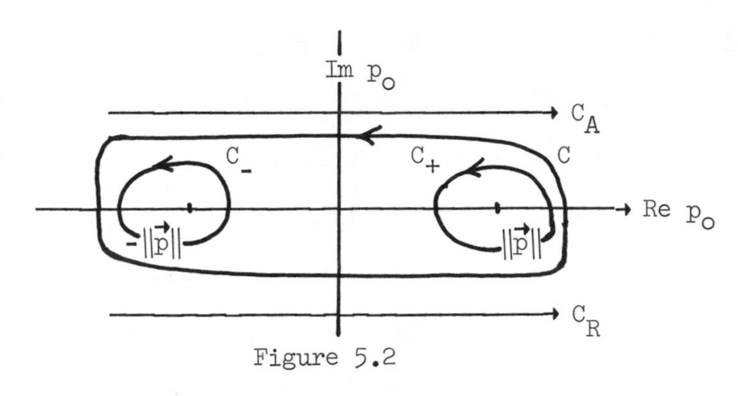

Figure 5.2

To obtain a fundamental solution when $x_o > 0$ (future, retarded), choose the contour C_R . As (5.32) is invariant under time inversion, $x_o \leftrightarrow -x_o$, there will be a solution appropriate to $x_o < 0$ (past, advanced) which arises from the contour C_A . Consequently, with the full use of relativistic notation we define

$$\gamma_{ret,\ adv} = -1/(2\pi)^{n+1} \int_{C_R, C_A} dp \; e^{ipx}/p^2 \quad . \tag{5.34}$$

If $x_o < 0$, closing the contour C_R in the lower half plane, yields zero for γ_{ret} . When $x_o > 0$, completing C_R in the upper half plane produces a contour which is equivalent to C . Similar remarks hold for γ_{adv} with the result

$$\gamma_{ret,adv} = \pm\, H(\pm\, x_o)\gamma(x) \quad . \tag{5.35}$$

One last relation between the contours appearing in Figure 5.2 arises by decomposing C into the two contours C_+ and C_- . If γ_+ and γ_- denote the expressions so obtained by replacing C in (5.33) by C_+ and C_- , respectively, we find

$$\gamma = \gamma_+ + \gamma_- = \gamma_{ret} - \gamma_{adv} \tag{5.36}$$

$\gamma_+(\gamma_-)$ is referred to as the positive (negative) energy part of γ since in the special theory of relativity a zero mass particle with space momentum $\vec{p}$ has energy given by $\|\vec{p}\|$.

The properties of the various fundamental solutions introduced above follow directly from those of the basic generalized function γ . Let us now examine these in detail. A further representation for γ , which allows us to exploit fully the properties of Fourier transforms, is found by evaluating the residues at the poles in the contours $C_\pm$ and using the relation in problem 2.2 (1) ; namely,

$$\gamma_\pm = (\mp i)/(2\pi)^n \int d\vec{p}\, e^{ipx}/2\|\vec{p}\| \qquad p_0 = \pm\|\vec{p}\|$$

$$= (\mp i)/(2\pi)^n \int_{R^{n+1}} dp_0\, d\vec{p}\, H(\pm p_0)\delta(p^2)e^{i\,px} \quad .$$

Combining these expressions as in (5.36) yields

$$\gamma(x) = -i/(2\pi)^n \int_{R^{n+1}} dp \text{ sgn } p_0\ \delta(p^2)\ e^{i\,px} \quad . \tag{5.37}$$

In this form γ is sometimes referred to as the Pauli-Jordan distribution by virtue of its systematic use in quantum electrodynamics. We summarize the essential properties of γ below.

PROPOSITION 5.1. The expression (5.37) defines γ as a tempered distribution which is invariant under $\mathcal{L}_+^\uparrow$, the proper time preserving Lorentz group. Moreover, as an element of $\mathcal{S}'(R^{n+1})$, γ satisfies

(a) $\Box_n \gamma = 0$ (b) $\gamma|_{x_0=0} = 0$

(c) $\partial\gamma/\partial x_0|_{x_0=0} = \delta(\vec{x})$ (d) $\operatorname{supp} \gamma \subset V_+ \cup V_-$.

Proof. The manipulations leading to (5.37) were carried out formally and may be justified in terms of our discussion in section (3.7). Disregarding numerical factors, γ is the Fourier transform of the distribution $\operatorname{sgn} p_0\ \delta(p^2)$. This is tempered, since for any $f \in \mathcal{S}(R^{n+1})$, the dual relation of Definition 3.12 gives

$$\gamma(f) = (2\pi)^{-n-1}\langle\tilde{\gamma}, \dot{\tilde{f}}\rangle = -i(2\pi)^{-2n-1}\int dp\, \operatorname{sgn} p_0\ \delta(p^2)\tilde{f}(-p)$$

$$= -i(2\pi)^{-2n-1}\int_{R^n} d\vec{p}\ g(\vec{p})$$

where $g(\vec{p}) = [f(-\|\vec{p}\|, \vec{p}) - f(\|\vec{p}\|, \vec{p})]/2\|\vec{p}\|$ is a test function in $\mathcal{S}(R^n)$.

For the various properties listed above, note

(a) $\Box_n \gamma$ is the Fourier transform of the distribution $p^2 \operatorname{sgn} p_0 \delta(p^2)$ which is zero as $\delta(p^2)$ has support on the light cone, $p^2 = 0$.

(b) γ restricted to the space hyperplane, $x_0 = 0$, is the Fourier transform of $\int dp_0\ \operatorname{sgn} p_0\ \delta(p^2)$ which is zero.

(c) a brief calculation shows that $\partial\gamma/\partial x_0$ for $x_0 = 0$ is the Fourier transform in the space variables of the function 1 .

(d) From (4.31), γ is $\mathcal{L}_+^\uparrow$ - invariant if, and only if, $\operatorname{sgn} p_0\ \delta(p^2)$ is $\mathcal{L}_+^\uparrow$ - invariant. This is clearly the case since Lorentz transformations in $\mathcal{L}_+^\uparrow$ do not reverse the sign of p_0 (see Lemma 4.19). In (b), we have shown that γ vanishes in the space hyperplane. Thus, γ is zero at all space-like points by virtue of its Lorentz invariance.

Proposition 5.1 indicates that γ is not a fundamental solution in the sense of (5.32). This is the case, however, for $\gamma_{ret,\ adv}$.

PROPOSITION 5.2. The expressions $\gamma_{ret,\ adv}$ defined by (5.34) are $\mathcal{L}_+^\uparrow$ invariant tempered distributions with the properties

(a) $\Box_n \gamma_{ret,\ adv} = \delta_x$ (b) $\gamma_{ret,\ adv}|_{x_o=0} = 0$

(c) $\text{supp}\ \gamma_{ret,\ adv} \subset V_\pm$.

Proof. The first property is most easily seen from the representation (5.34), whereby $\Box_n \gamma_{ret,\ adv}$ becomes the Fourier transform of 1 as the wave operator $\Box_n$ removes the singularities in the integrand and $C_{R,A}$ are both equivalent to the contour R^{n+1} . The remaining properties follow from Proposition 5.1 and (5.35).

The relation (5.35) indicates that γ is the correct choice of fundamental solution for the Cauchy problem on R^n with similar results for either positive or negative time, x_o .

THEOREM 5.3. Consider the equation $\Box_n \phi = 0$ with Cauchy data

(i) $\phi|_{x_o=0} = \alpha$ (ii) $\partial\phi / \partial x_o|_{x_o=0} = \beta$

in which α and β are in $\mathcal{S}'(R^n)$. Then $\phi \in \mathcal{S}'(R^{n+1})$, and is given by the relation

$$\phi(x) = \int_{y_0=0,\, \vec{y}\in R^n} d\vec{y}\,[\gamma(x-y)\beta(\vec{y}) - \partial\gamma(x-y)/\partial y_0\; \alpha(\vec{y})]$$

<u>Proof</u>. Let us deal first with the regularity of ϕ. Rewrite the first term in the form $\gamma_t * \beta$ where $\gamma_t = \gamma(t, \vec{x})$. By (d) in Proposition 5.1, for a fixed value of t, γ_t has support in the ball $\|\vec{x}\| \leq t$ which is compact. This implies for $\beta \in \mathcal{S}'(R^n)$ that $\gamma_t * \beta$ is tempered by Lemma 2.24. Similar remarks hold for the second term.

Interpreting the two terms as convolution with γ_t, we see that ϕ is a solution of the homogeneous wave equation by (a) of Proposition 5.1. That ϕ has the correct initial data is a consequence of (b) and (c) in this proposition; that is,

$$\lim_{t\to 0} \phi(t, \vec{x}) = \lim_{t\to 0} \partial(\gamma_t * \alpha)/\partial t = \delta * \alpha = \alpha$$

$$\lim_{t\to 0} \partial\phi(t, x)/\partial t = \lim_{t\to 0} \partial(\gamma_t * \beta)/\partial t = \beta$$

Each of these limits is taken in the $\mathcal{S}'(R^n)$ topology and in the second line we have made use of the fact $\partial^2\gamma/\partial t^2|_{x_0=0} = \Delta_n\gamma|_{x_0=0} = 0$. This may be derived in the same manner as for (c) in Proposition 5.1.

PROBLEMS (1) Let $\Lambda \in \mathcal{L}_+^\uparrow$ and consider $x' = \Lambda x$. Show that if γ is a solution to (5.32), then so is $\Lambda\gamma$. Hence, γ is Lorentz invariant.

(2) In Theorem (5.3), use the dual relation of Definition 3.12 to show that for $f \in \mathcal{S}(R^{n+1})$, $\phi(f) = \beta(g) + \alpha(h)$ where $g = -i(2\pi)^{-2n-1}[f(-\|\vec{p}\|, \vec{p}) - f(\|\vec{p}\|, \vec{p})]/2\|\vec{p}\|$ and $h = -(1/2)(2\pi)^{-2n-1}[f(-\|\vec{p}\|, \vec{p}) + f(\|\vec{p}\|, \vec{p})]$ are test functions in $\mathcal{S}(R^n)$.

(3) Repeat the discussion of this section for the wave equation with positive mass, (5.31), and show that the same results hold with appropriate modifications. You will find it useful when dealing with the support properties to define $V_{m,+} = \{p \in R^{n+1} | p^2 \geq m^2 , p_0 > 0\}$.

Before discussing some of the particular features of the solution to the wave equation obtained in Theorem 5.3, let us turn our attention to the inhomogeneous equation. In this case, γ_{ret} will be the appropriate fundamental solution.

THEOREM 5.4. Let $\psi \in \mathcal{S}'(R^{n+1})$ be zero for $x_0 < 0$. Then the equation $\Box_n \phi = \psi$ with Cauchy data

(i) $\phi|_{x_0=0} = 0$ (ii) $\partial\phi/\partial x_0|_{x_0=0} = 0$

has solution $\phi_{ret} = \gamma_{ret} * \psi$. ϕ_{rct} is called a retarded potential.

Proof. Consider a fixed space-time point x. The support property of γ_{ret} implies that only those values of ψ contribute to $\gamma_{ret} * \psi$ which lie in the backward light cone of x above the space hyperplane. This region is compact and $\psi \in \mathcal{S}'$ implies $\phi_{ret} \in \mathcal{S}'$ (this is an application of Lemma 2.25 where the supports of both γ_{ret} and ψ are bounded below). This takes care of the regularity properties of ϕ_{ret}. The remaining properties follow easily from those of Proposition 5.2.

A more general set of initial conditions for the inhomogeneous equation may be obtained by adding a suitable solution of the homogeneous equation in the form given by Theorem 5.3.

5.5 EXAMPLES FOR THE WAVE OPERATOR

The solutions to the wave equation found in Theorems 5.3 and 5.4 exhibit several rather interesting features regarding the propagation of the initial data by means of the wave operator. We collect these in this section together with some detailed calculations of the fundamental solutions. These latter results are peculiar to the zero mass equation.

DOMAIN OF DEPENDENCE. Consider the homogeneous wave equation appearing in Theorem 5.3. Given a point $(t, \vec{x})$ in space-time, we ask for the values of the Cauchy data which determine the solution ϕ at this point. Since both γ and $\partial\gamma/\partial x_0$ have support in $V_+ \cup V_-$, only those values of the data in R^n which lie in the backward cone of $(t, \vec{x})$ have any effect. This region is called the domain of dependence for x at time t (for $t > 0$ see Figure 5.3).

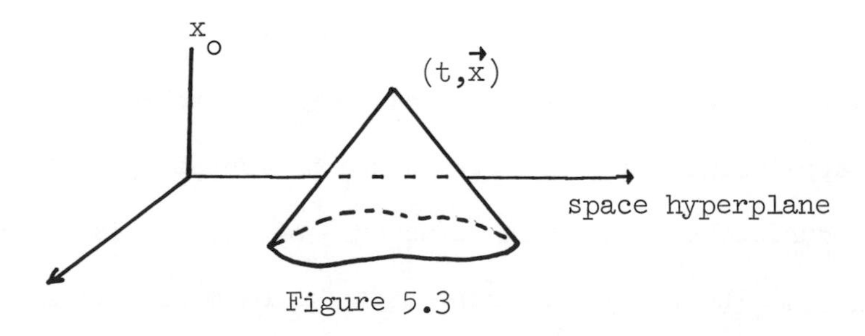

Figure 5.3

A similar region results if $t < 0$.

DOMAIN OF INFLUENCE. Suppose $\beta = 0$ and $\alpha \in C^{\infty\,\prime}$ in Theorem 5.3. Let us pose the converse question to the above; namely, for what values of $(t, \vec{x})$, $t > 0$, does the solution, ϕ, depend on α. Clearly, only those points $\vec{x}$ need be considered for which the backward cone at x intersects supp α. The region influenced by α is then envelopped by the outgoing light rays from the boundary of supp α (see Figure 5.4).

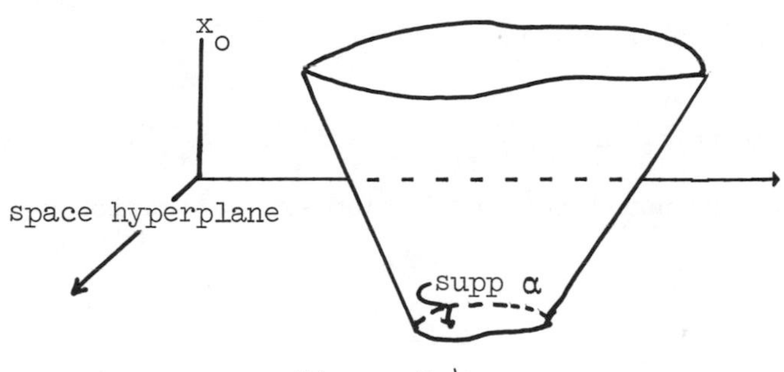

Figure 5.4

FINITE PROPAGATION SPEED. Suppose that the Cauchy data is a discrete disturbance at $t = 0$ at the origin. The domain of influence of this signal for positive time will be the forward light cone and its interior. Consequently, the effect of the signal at a point x is not measured until a time $t = \|\vec{x}\|$ has elapsed. The wave operator propagates the initial data with a finite velocity, which is one in our units, in contrast to the heat operator where propagation is instantaneous. The same observer would also detect the signal as radiating spherically from the origin. For the zero mass wave operator in an odd number of space dimensions greater than one, there is a sharper characterization for these spherical waves due to Huyghens. This is revealed below.

FUNDAMENTAL SOLUTIONS FOR n ODD. We next obtain explicit forms for γ, or $\gamma_{ret,adv}$ when the number of space dimensions is odd. The simplest case is that of one dimension whereupon from (5.37) and (3.22) we find

$$\gamma = 1/(2\pi i)\int_{-\infty}^{\infty} dp\left[e^{i(|p|x_o-px)} - e^{-i(|p|x_o+px)}\right]/2|p|$$

$$= 1/\pi\int_{-\infty}^{\infty} dp \sin(|p|x_o)\cos(p|x|)/|p|$$

$$= (1/4)\left[\operatorname{sgn}(x_o+|x|)+\operatorname{sgn}(x_o-|x|)\right] .$$

It follows from this, that $\gamma_{ret,adv} = (1/2)H(x_o \mp |x|)$, and Theorem 5.3 reduces to the usual form of D'Alembert's solution to the one dimensional wave equation

$$\phi(t,x) = (1/2)[\alpha(x+t)+\alpha(x-t)]+1/2\int_{x-t}^{x+t} dy\ \beta(y) . \qquad (5.38)$$

The general case for n-dimensions is not much more complicated. It helps to keep in mind that the final result must be Lorentz invariant even though this is not explicit during the calculation. Again, from (5.37), we obtain

$$\gamma(x) = 1/(2\pi)^n\int_{R^n} d\vec{p}\ \sin(\|\vec{p}\|\ x_o)/\|\vec{p}\|\ e^{-i\ \vec{p}\cdot\vec{x}} .$$

Choose spherical coordinates as in problem 4.3 (8) with $\vec{p}\cdot\vec{x} = \|\vec{p}\|\ \|\vec{x}\| \cos\theta_1$ and change the angular variable θ_1 to $z = \cos\theta_1$. After a brief calculation, this expression becomes

$$\gamma(x) = \frac{(-1)^{(n-1)/2}\Omega_{n-1}}{(2\pi)^n\|\vec{x}\|} \frac{\partial^{n-2}}{\partial x_o^{n-2}} \int_0^\infty dr \cos(r\, x_o/\|\vec{x}\|) \cdot$$

$$\int_{-1}^{+1} dz\ (1-z^2)^{(n-3)/2}\ e^{-irx}$$

$$= \frac{(-1)^{(n-1)/2}\Omega_{n-1}}{2(2\pi)^n\|\vec{x}\|} \frac{\partial^{n-2}}{\partial x_o^{n-2}} \int_{-1}^{+1} dz(1-z^2)^{(n-3)/2} \int_{-\infty}^{\infty} dr\, e^{ir(x_o/\|\vec{x}\|-z)} .$$

After substituting (3.23) and simplifying, the last integral yields

$$\gamma_{ret,adv} = \frac{\pi\,\Omega_{n-1}}{(2\pi)^n\|\vec{x}\|^{n-2}} \frac{\partial^{n-2}}{\partial x_o^{n-2}} (x^2)^{(n-3)/2} H(x_o \mp \|\vec{x}\|) \quad . \qquad (5.39)$$

An important special case is when $n = 3$. Then (5.39) takes the particularly simple form

$$\gamma_{ret,adv} = (1/4\pi\|\vec{x}\|)\ \delta(x_o \mp \|\vec{x}\|) \quad . \qquad (5.40)$$

Inserting (5.40) into both Theorems 5.3 and 5.4 leads to the expressions $(t > 0)$

$$\phi(t,\vec{x}) = \int_{\|\vec{x}-\vec{y}\|=t} d\vec{y}\ \beta(\vec{y})/(4\pi t) + \partial/\partial t \int_{\|\vec{x}-\vec{y}\|=t} d\vec{y}\ \alpha(\vec{y})/(4\pi t) \qquad (5.41)$$

$$\phi_{ret}(t,\vec{x}) = \int_{\|\vec{x}-\vec{y}\|\le t} d\vec{y}\ \psi(t - \|\vec{x}-\vec{y}\|,\ \vec{y})/(4\pi\|\vec{x}-\vec{y}\|) \quad .$$

Let us examine the domain of dependence for the initial data in (5.41) more closely. In the space plane, the Cauchy data is actually concentrated on the sphere $\|\vec{x}-\vec{y}\| = t$. Thus, an observer at $\vec{x}$ after a time t detects a radiating signal which moves past in a spherical wave front. After the time t , there is no longer any effect. The propagation of a signal by means of these sharp wave fronts is sometimes referred to as Huyghens' Principle and is a characteristic feature of the wave operator for an odd number of space dimensions. The general case follows from (5.39) by noting that there will always be one time derivative to convert the step function into a δ-distribution. Huyghens' Principle is a consequence of the sharp support of this distribution. For one or an even number of space dimensions, this is no longer true. The signal propagates shperically, but diffuses in the sense that after an elapsed time t, an observer at $\vec{x}$ still measures a non-zero wave field ϕ. This may be verified for one dimension by using (5.38). For general $n = 2s$, the method of descent, [11], allows one to obtain a solution by using the expression (5.39) for $n = 2s + 1$.

BIBLIOGRAPHY

The following list is selected from the extensive literature on the subjects of Fourier analysis, distributions and partial differential equations and represents works suitable for further study. References [1]to [5] pertain to Chapters one, two and three; references [7],[8] and [9] deal with the beginning of Chapter three. For Chapter five, references [11]and [12] discuss the classical theory of partial differential equations, while [13] is an account of the modern work. On the subject of singular integrals, [14] describes the classical and recent results. In reference [6], the reader will find a comprehensive account of all of these subjects within the framework of functional analysis. For those who wish to carry further the relation between spaces of sequences and tempered distributions introduced in Chapter three, references [15] and [16] provide interesting and detailed accounts.

1. J. L. Kelley, <u>General Topology</u>, D. Van Nostrand, Princeton, 1955.

2. I. M. Gelfand and G. E. Shilov, <u>Les Distributions</u>, Tome 2, Dunod, Paris, 1964.

3. L. Schwartz, Théorie des Distributions, Parts I and II, Hermann, Paris, 1957 and 1965.

4. F. Treves, Topological Vector Spaces, Distributions, and Kernels, Academic Press, New York, 1967.

5. H. Bremermann, Distributions, Complex Variables, and Fourier Transforms, Addison-Wesley, Reading, 1965.

6. K. Yosida, Functional Analysis, Academic Press, New York, and Springer-Verlag, Berlin, 1965.

7. R. Seeley, An Introduction to Fourier Series and Integrals, W. A. Benjamin, New York, 1966.

8. E. T. Whittaker and G. N. Watson, Modern Analysis, 4 th. Edition, University Press, Cambridge, 1958.

9. G. H. Hardy, Pure Mathematics, 10 th Edition, University Press, Cambridge, 1955.

10. P. D. Methee, Comm. Math. Helv., 28, 225 (1954).

11. R. Courant and D. Hilbert, Methods of Mathematical Physics, 2 Vols., Interscience, New York, 1953 and 1966.

12. G. Hellwig, Partial Differential Equations, Blaisdell, New York, 1964.

13. L. Hormander, Linear Partial Differential Operators, Academic Press, New York, and Springer-Verlag, Berlin, 1963.

14. E. M. Stein, Singular Integrals and Differentiability Properties of Functions, Princeton University Press, Princeton, 1970.

15. B. Simon, Distributions and their Hermite Expansions, Journal of Mathematical Physics, 12, 140 (1971).

16. V. Bargmann, Hilbert Spaces of Analytic Functions, Part II, Comm. on Pure and Applied Mathematics, 20, 1 (1967).

INDEX